Water Supplies For Fire Protection

Fourth Edition

VALIDATED BY
THE INTERNATIONAL FIRE SERVICE TRAINING ASSOCIATION

PUBLISHED BY
FIRE PROTECTION PUBLICATIONS
OKLAHOMA STATE UNIVERSITY

Cover Photo Courtesy of:
Bob Esposito

Dedication

This manual is dedicated to the members of that unselfish organization of men and women who hold devotion to duty above personal risk, who count on sincerity of service above personal comfort and convenience, who strive unceasingly to find better ways of protecting the lives, homes and property of their fellow citizens from the ravages of fire and other disasters . . . The Firefighters of All Nations.

Dear Firefighter:

The International Fire Service Training Association (IFSTA) is an organization that exists for the purpose of serving firefighters' training needs. Fire Protection Publications is the publisher of IFSTA materials. Fire Protection Publications staff members participate in the National Fire Protection Association and the International Association of Fire Chiefs.

If you need additional information concerning our organization or assistance with manual orders, contact:

Customer Services
Fire Protection Publications
Oklahoma State University
930 N. Willis
Stillwater, OK 74078-8045
1 (800) 654-4055

For assistance with training materials, recommended material for inclusion in a manual, or questions on manual content, contact:

Technical Services
Fire Protection Publications
Oklahoma State University
930 N. Willis
Stillwater, OK 74078-8045
(405) 744-5723

First Printing, February 1988
Second Printing, May 1990
Third Printing, August 1993
Fourth Printing, August 1998
Fifth Printing, July 1999

ISBN 0-87939-073-5
Library of Congress 87-83172
Fourth Edition
Printed in the United States of America

Table of Contents

List of Tables

THE INTERNATIONAL FIRE SERVICE TRAINING ASSOCIATION

The International Fire Service Training Association is an educational alliance organized to develop training material for the fire service. The annual meeting of its membership consists of a workshop conference which has several objectives —

. . . to develop training material for publication
. . . to validate training material for publication
. . . to check proposed rough drafts for errors
. . . to add new techniques and developments
. . . to delete obsolete and outmoded methods
. . . to upgrade the fire service through training

This training association was formed in November 1934, when the Western Actuarial Bureau sponsored a conference in Kansas City, Missouri, to determine how all agencies that were interested in publishing fire service training material could coordinate their efforts. Four states were represented at this conference and it was decided that, since the representatives from Oklahoma had done some pioneering in fire training manual development, other interested states should join forces with them. This merger made it possible to develop nationally recognized training material which was broader in scope than material published by an individual state agency. This merger further made possible a reduction in publication costs, since it enabled each state to benefit from the economy of relatively large printing orders. These savings would not be possible if each individual state developed and published its own training material.

From the original four states, the adoption list has grown to forty-four American States; six Canadian Provinces; the British Territory of Bermuda; the Australian State of Queensland; the International Civil Aviation Organization Training Centre in Beirut, Lebanon; the Department of National Defence of Canada; the Department of the Army of the United States; the Department of the Navy of the United States; the United States Air Force; the United States Bureau of Indian Affairs; The United States General Services Administration; and the National Aeronautics and Space Administration (NASA). Representatives from the various adopting agencies serve as a voluntary group of individuals who govern policies, recommend procedures, and validate material before it is published. Most of the representatives are members of other international fire protection organizations and this meeting brings together individuals from several related and allied fields, such as:

. . . key fire department executives and drillmasters,
. . . educators from colleges and universities,
. . . representatives from governmental agencies,
. . . delegates of firefighter associations and organizations, and
. . . engineers from the fire insurance industry.

This unique feature provides a close relationship between the International Fire Service Training Association and other fire protection agencies, which helps to correlate the efforts of all concerned.

The publications of the International Fire Service Training Association are compatible with the National Fire Protection Association's Standard 1001, "Fire Fighter Professional Qualifications (1981)," and the International Association of Fire Fighters/International Association of Fire Chiefs "National Apprenticeship and Training Standards for the Fire Fighter." The standards are an effort to attain professional status through progressive training. The NFPA and IAFF/IAFC Standards were prepared in cooperation with the Joint Council of National Fire Service Organizations of which IFSTA is a member.

The International Fire Service Training Association meets each July at Oklahoma State University, Stillwater, Oklahoma. Fire Protection Publications at Oklahoma State University publishes all IFSTA training manuals and texts. This department is responsible to the executive board of the association. While most of the IFSTA training manuals can be used for self-instruction, they are best suited to group work under a qualified instructor.

Preface

The importance of an adequate and reliable water supply cannot be overemphasized during fire fighting operations. Knowledge of the most efficient ways to use the available water supply system is also critical for effective and efficient fireground operations. In areas that do not have an organized water distribution system, firefighters have the additional task of drafting and relaying water to the fire scene. In these cases, knowledge of suitable emergency water sources is essential.

This manual is designed to acquaint fire officials with water supply management; water supply fundamentals; types, installation, distribution, and inspection of fire hydrants; fire flow testing; static sources; relay operations; and various other aspects of water supplies for the fire service.

Acknowledgements and grateful thanks are extended to the validating committee, who assisted with the final draft of this manual:

Chairman
James Simms
Rolf Jensen and Associates
Pleasant Hill, California

Don Beavin
ISO Commercial Risk Services
Oklahoma City, Oklahoma

Ronald Callahan
Indiana Fire Instructors Assn.
Speedway, Indiana

Bob Caron
Fire Instructors Assn. of Minn.
Detroit Lakes, Minnesota

Eric Haussermann
Oklahoma Fire Service Training
Stillwater, Oklahoma

Jesse Jackson
Maryland Fire & Rescue Institute
College Park, Maryland

Max McRae
Houston Fire Department
Houston, Texas

Robert Noll
Yukon Fire Department
Yukon, Oklahoma

Luther Sharbono
Shreveport Fire Department
Shreveport, Louisiana

Russell Strickland
Maryland Fire & Rescue Institute
College Park, Maryland

To the other members who contributed to the committee during its tenure, we also express our appreciation:

William J. Matheny
Fire Education, Inc.
Tucker, Georgia

Frank McGurn
National Fire Sprinkler Assn., Inc.
Crystal Lake, Illinois

Edward Steiner
OSU Fire Service Training
Pauls Valley, Oklahoma

Gratitude is also extended to the following individuals and organizations for supplying us with needed information, photographs, and illustrations for this manual:

Pennsburg Fire Co. No. 1
    Chief Engineer Larry Buck
    Firefighter/Photographer Bob Esposito
Edmond, OK Fire Department
City of Stillwater, OK Water Department
City of Stillwater Fire Services
    Captain Lanny Porter
Elkhart Brass Manufacturing Co., Inc.
Ziamatic Corp.
Jaffery Fire Protection Co., Inc.
Joel Woods, Maryland Fire & Rescue Institute
Insurance Services Offices Commercial Risk Services, Inc.
Ken-Mar, Inc.
David Grupp
Marv Ackerman
Tower Fire Apparatus
Merced City, CA Fire Department
Monroe Truck Equipment, Inc.
United States Coast Guard
Clinton H. Smoke, Jr.
Richard Mahaney
Oklahoma State University Safety Department
    Richard Giles, Director
    Michael D. Lunsford, Safety Specialist

Special thanks are also due to William Eckman of the Maryland Fire and Rescue Institute, whose considerable knowledge and many years of teaching on the subject of water supplies are reflected in several sections of this manual.

Finally, gratitude is extended to the following members of the Fire Protection Publications staff who made the final publication of this manual possible:

Carol Smith, Publications Specialist
Michael A. Wieder, Associate Editor
Lynne Murnane, Senior Publications Editor
Robert Fleischner, Publications Specialist
Don Davis, Production Coordinator
Ann Moffat, Graphic Designer
Laurie Zirkle, Graphic Artist
Mike McDonald, Graphic Artist
Desa Porter, Phototypesetter Operator II
Karen Murphy, Phototypesetter Operator II
Cindy Brakhage, Unit Assistant
Terri Jo Gaines, Senior Clerk/Typist
Scott Tyler, Research Technician
Brett Lacey, Research Technician

Gene P. Carlson
Managing Editor

Glossary

A

ALTITUDE — The geographical position of a location or object in relation to sea level. The location may be either above, below, or at sea level.

ATMOSPHERIC PRESSURE — The pressure exerted by the atmosphere at the surface of the earth due to the weight of air. Atmospheric pressure at sea level is about 14.7 psi (101 kPa); it increases as elevation decreases below sea level and decreases as elevation increases above sea level.

AVERAGE DAILY CONSUMPTION — The average of the total amount of water used each day during a one-year period.

B

BACKDRAFT — Instantaneous combustion that occurs when oxygen is introduced into a smoldering fire. The stalled combustion resumes with explosive force.

BOILING POINT — The temperature at which the vapor pressure of a liquid is equal to the external pressure applied to it.

BRITISH THERMAL UNIT (BTU) — The amount of heat energy required to raise the temperature of one pound (0.45 kg) of water one degree Fahrenheit. One BTU = 1.055 kj.

C

C-FACTOR — A factor that indicates the roughness of the inner surface of the piping. The C-factor decreases as the sediment, incrustation, and tuberculation within the pipe increases.

CALORIE — The amount of heat needed to raise the temperature of one gram of water one degree Centigrade.

CAPACITY — The maximum ability of a pump or water distribution system to deliver water.

CENTIGRADE (CELSIUS) — Temperature scale on which O° (32°F) is the freezing point of water and 100° (212°F) is the boiling point of water.

CIRCLE OR BELT SYSTEM — A water supply system designed in the form of a large loop. Water may be supplied to any part of the system from two directions, resulting in less overall friction loss.

CIRCULATING SYSTEM — See circle or belt system.

CISTERN — A water storage receptacle that is usually underground and supplied by a well.

COEFFICIENT OF DISCHARGE — A correction factor, relating to the shape of the hydrant discharge outlet, used when computing the flow from a hydrant.

COMBINATION ATTACK — Using both a direct and an indirect attack on a fire. This method utilizes the steam generating technique of a ceiling level attack along with an attack on the burning materials near floor level.

COMBINATION SYSTEM — A water supply system that is a combination of a gravity system and a direct pumping system. It is the most common type of municipal water supply system.

COMBUSTION—A rapid oxidation of a fuel by a chemical reaction that is self-sustaining and exothermic. In addition to heat, light, gases, and other particles are given off.

5-90

COMMUNITY MASTER PLAN — A medium- to long-range plan of goals for the planned, orderly growth and development of a particular community. The plan is generally written through the interaction of all city agencies that may be affected by future development.

CONDUCTION — The transfer of heat energy from one body to another through a solid medium.

CONSTANT PRESSURE RELAY — A method of establishing a relay water supply utilizing two or more pumpers to supply the attack pumper. This method reduces the need for time consuming and often confusing fireground calculations of friction loss.

CONVECTION — The transfer of heat energy by the movement of air or liquid.

D

DEFENSIVE ATTACK — An exterior fire attack used when the known fire flow is not sufficient for a given fire, or it is obvious that the building will be lost regardless of the available water flow.

DIKE — A temporary dam constructed of readily available objects used to obstruct the flow of a shallow stream of water to a depth that facilitates drafting operations.

DIRECT ATTACK — The application of a fire stream directly onto a burning fuel.

DIRECT PUMPING SYSTEM — A water supply system supplied directly by a system of pumps, rather than elevated storage tanks.

5-90

DISPLACEMENT—The amount of water drawn into the pump, thus displacing air. Also, the volume of water displaced by a stroke of a positive displacement pump.

DOMESTIC CONSUMPTION — Water consumed from the water supply system by residential and commercial occupancies.

DRAFT — To draw water from a static source into a pump above the level of the water source.

DRY-BARREL HYDRANT — A type of fire hydrant that has its operating valve right at the water main, rather than in the barrel of the hydrant itself. In this way, no water is in the barrel of the hydrant when it is not in use. Dry barrel hydrants are used in areas where freezing is likely to occur.

DRY HYDRANT — A permanently installed pipe that has pumper suction connections installed at static water sources to speed drafting operations.

E

ELEVATED STORAGE — A water storage reservoir located well above the level of the system it supplies; the elevation takes advantage of head pressure to supply the system.

5-90

ENDOTHERMIC REACTION—A chemical reaction in which a substance absorbs heat energy.

EXOTHERMIC REACTION—A chemical reaction in which heat energy is released.

EXPLOSIVE LIMIT — See Flammable Limit.

EXTERIOR EXPOSURE — A building or other combustible object located close to the fire building that is in danger of becoming involved due to heat transfer from the fire building.

F

FAHRENHEIT — A temperature scale on which 32° (0°C) is the freezing point of water and 212° (100°C) is the boiling point of water.

FILL SITE — The location at which tankers are loaded during a water shuttle operation.

FIRE DEPARTMENT WATER SUPPLY OFFICER — The officer in charge of all water supplies at the scene of a fire. Duties include placing pumpers at most advantageous hydrants or other water sources and direction of supplementary water supplies, such as water shuttles and relay pumping operations. This may also be a full-time staff position in which the person coordinates water supply projects of concern to the fire department with other agencies involved in the water supply system.

FIRE FLOW — The quantity of water available for fire fighting in a given area; it is calculated in addition to the normal water consumption in the area.

FIRE FLOW TESTING — The procedure used to determine the rate of water flow available for fire fighting at various points within the distribution system.

FIREGROUND COMMANDER — The officer in charge at a fire or other emergency handled by the fire department.

FIRE POINT — The temperature at which a liquid fuel produces sufficient vapors to support combustion once the fuel is ignited. The fire point is usually a few degrees above the flash point.

FIRE RESISTIVE — The ability of a structure or material to provide a predetermined degree of fire resistance as required by building and fire prevention codes. Fire resistance is given in hour ratings or a fraction thereof.

FIRE STREAM — A stream of water or other water-based extinguishing agent after it discharges from the nozzle until it reaches the desired point.

5-90

FLAMMABLE LIMIT—The range of concentration of a substance in air that can be ignited. Most substances have an upper (too rich) and lower (too lean) flammable limit.

FLASHOVER — The stage of a fire at which all surfaces and objects are heated to their ignition temperature and flame breaks out almost at once over the entire surface.

FLASH POINT — The minimum temperature at which a liquid fuel gives off sufficient vapors to form an ignitable mixture with the air near the surface. At this temperature, the ignited vapors will flash but will not continue to burn.

FLOW — The motion characteristic of water.

FLOW PRESSURE — Pressure created by the rate of flow or velocity of water coming from a discharge opening.

FORCE — The simple measure of weight, usually expressed in pounds or kilograms.

FRICTION LOSS (PRESSURE LOSS DUE TO FRICTION) — That part of the total pressure lost as water moves through a hose or piping system, caused by water turbulence and the roughness of interior surfaces of hose or pipe.

G

GRAVITY SYSTEM — A water supply system which relies entirely on the force of gravity to propel the water throughout the system. This type of system is generally used in conjunction with an elevated water storage source.

GRID SYSTEM — A water supply system which utilizes lateral feeders for improved distribution.

H

5-90

HEAD PRESSURE—That pressure exerted by a stationary column of water, directly proportional to the height of the column.

HARD SUCTION (HARD SLEEVE) — A somewhat flexible, rubberized length of noncollapsible hose with a steel core that connects the pump to a water source. It is most commonly used for drafting operations.

HEAT — The form of energy that raises temperature. Heat is measured by the amount of work it does.

HYDRANT PRESSURE — The amount of pressure being supplied by a hydrant without assistance.

I

IGNITION TEMPERATURE — The minimum temperature to which a fuel in air must be heated in order to start a self-sustained combustion reaction independent of the heating source.

IMPINGING STREAM NOZZLE — A nozzle that drives several jets of water together at a set angle for the purpose of breaking the water into finely divided particles.

IMPLOSION — The rapid inward collapsing of the walls of a vessel or structure because of the inability of the walls to sustain a vacuum.

IMPOUNDED WATER SUPPLY — Generally used to describe an open, standing, man-made reservoir, but can be used to describe any type of standing, static water supply.

INDIRECT ATTACK — The direction of a fire stream at the ceiling level of a room or building in order to generate a large amount of steam. The steam helps darken the fire and cools the area so that firefighters may enter. At this point, they can make a direct attack to totally extinguish the fire.

INDUSTRIAL CONSUMPTION — That water consumed from the water supply system by industrial facilities.

INSOLUBLE — Incapable of being dissolved in a liquid (usually water).

INTERIOR EXPOSURE — Areas of a fire building that are not involved in fire, but that are connected to the fire area in such a manner that may facilitate fire spread through any available openings.

J

JET SIPHON — A section of pipe or hard suction hose with a 1½-inch (38 mm) discharge line inside that bolsters the flow of water through the tube. The jet siphon is used between portable tanks to maintain a maximum amount of water in the tank from which the pumper is drafting.

L

LATENT HEAT OF VAPORIZATION — The quantity of heat absorbed by a substance when it changes from a liquid to a vapor.

LOCATION MARKER — A device such as a reflective marker or flag used to mark the location of a fire hydrant for quicker identification during a fire response.

M

MAXIMUM DAILY CONSUMPTION — The maximum total amount of water used during any 24-hour interval in a 3-year period.

MUTUAL AID — Two-way assistance by fire departments, often from adjoining jurisdictions. Such an agreement becomes very important when a fire is beyond the capabilities of the fire department whose jurisdiction it is within.

N

NET PUMP DISCHARGE (ENGINE) PRESSURE — The amount of pressure actually being created by the pump. In mathematical terms, it is the Pump Discharge Pressure minus the Pump Intake Pressure (PDP - PIP = NPDP).

NOMOGRAPH — A special chart, based on the Hazen-Williams formula, which can be used to assist in the determination of fire flows.

NORMAL OPERATING PRESSURE — That pressure found on a water distribution system during normal consumption demands.

NURSE TANKER — A very large water tanker (generally 4,000 gallons [15 142 L] or larger) that is stationed at the fire scene and serves as a portable reservoir, rather than as a shuttle tanker.

O

OFFENSIVE ATTACK — Usually an interior attack aimed directly at the seat of the fire. The needed fire flow must be available to initiate this attack.

OXIDATION — A chemical reaction in which oxygen combines with other substances. Fire, explosions, and rusting are examples of oxidation reactions.

P

PEAK HOURLY CONSUMPTION — The maximum amount of water used in any given hour of a day.

PERIPHERY-DEFLECTED FIRE STREAMS — Fire steams produced by deflecting water from the periphery of an inside circular stem in a fog nozzle against the exterior barrel of the nozzle.

PITOT TUBE — An instrument used to measure the velocity of a flowing fluid by converting the velocity energy to pressure energy which can then be measured by a pressure gauge.

5-90

PORTABLE PUMP — A small fire pump, available in several volume and pressure ratings, that can be removed from the apparatus and taken to a water supply inaccessible to the main pumper.

PORTABLE SOURCE — A source of water that is mobile and may be taken directly to the location where it is needed. This may be a fire department tanker or some other vehicle that is capable of hauling a large quantity of water.

PORTABLE TANK (RESERVOIR) — Any one of several styles of lightweight, easily set-up tanks used as a basin from which to draft when no other source is available. The tank is generally supplied by a shuttle of tankers which dump their water into it.

POUNDS PER SQUARE INCH (PSI) — The U.S. unit for measuring pressure. Its metric equivalent is kilopascals.

PRE-INCIDENT PLAN — A specific plan for fire fighting operations at a specific property or location in advance of any emergency situation.

PRESSURE — Force per unit area measured in pounds per square inch (kilopascals).

PRESSURE TANK — A water storage receptacle that uses compressd air pressure to propel water into the distribution system. Pressure tanks are generally small and will provide a very limited amount of water for fire protection.

PRIVATE CONNECTIONS — Connections to water supplies other than the standard municipal water supply system. These may include connection within a large industrial facility, a farm, or a private housing development.

PROPAGATION — The spread of combustion through a solid, gas, or vapor; the spread of fire from one combustible to another.

PUMP DISCHARGE PRESSURE — The total amount of pressure being discharged by a pump. In mathematical terms, it is the Pump Intake Pressure plus the Net Pump Discharge Pressure (PIP + NPDP = PDP).

PYROLYSIS — Chemical decomposition caused by heat that generally results in the lowered ignition temperature of a material.

R

RADIATION — The transfer of heat energy through light by electromagnetic waves.

RELAY OPERATION — Using two or more pumpers to move water over a long distance by operating them in series. Water discharged from one pumper flows through hoses to the inlet of the next pumper, and so on.

RELAY VALVE — A pressure relief device on the supply side of the pump designed to protect the hose and pump from damaging pressure surges common in relay pumping operations.

RESIDUAL PRESSURE — That part of the total pressure that is not used to overcome friction or gravity while forcing water through fire hose, pipe, fittings, and adapters.

S

SIPHON — A section of hard suction hose or piece of pipe to maintain an equal level of water in two or more portable tanks.

SOLUBILITY — The degree to which a solid, liquid, or gas dissolves in a solvent, usually water.

SOLUBLE — Capable of being dissolved in a liquid (usually water).

SPECIFIC HEAT — The ratio between the amount of heat required to raise the temperature of a specified quantity of a material and the amount of heat necessary to raise the temperature of an identical amount of water by the same number of degrees.

STAGING AREA — An area to which apparatus and personnel report, away from the emergency scene, to receive assignments at emergencies where there is an overall strategy.

STANDARD OPERATING PROCEDURE (S.O.P.) — Preplanned, written procedures by which the fire department operates under normal conditions. The principal benefit is that all companies in the department operate in basically the same manner, thus eliminating confusion when multiple companies are working together.

STATIC PRESSURE — Stored or potential energy that is available to force water through pipes and fittings, fire hose, and adapters.

STATIC SOURCE — A body of water that is not under pressure or in a supply piping system and must be drafted from to be used. Static sources include ponds, lakes, rivers, wells, and so on.

SUPPLEMENTAL PUMPING — Used when large fires overwhelm the water supply system. Supplemental pumping involves pumping water from a stronger point in the water system to the units at the fire or pumping it back into the water system in the area where the weakness is a problem.

T

TANKER (TENDER) — A mobile water supply fire apparatus that carries at least 1,500 gallons (5 678 L) of water and is used to supply water to fire scenes that lack fire hydrants.

TANKER SHUTTLE OPERATION — An operation where tankers deliver water to a fire scene, generally in a rotating order.

TENDER — See Tanker.

THREAD GAUGE DEVICE — A device that is screwed onto a hydrant discharge to check the condition of the threads and to ensure they are not damaged.

TREE SYSTEMS — A type of water supply piping system that utilizes a single, central feeder main to supply branches on either side of the main.

U

ULTIMATE CAPACITY — The total capacity of a water supply system including residential consumption, industrial consumption, available fire flow, and all other taxes on the system.

UNLOADING SITE — The point in the tanker shuttle operation where the portable tanks are located and the tankers unload their water.

V

VACUUM — A space completely devoid of matter or pressure. In fire service terms, it is more commonly used to describe a pressure that is somewhat less than atmospheric pressure. A vacuum is needed to facilitate drafting of water from a static source.

VAPORIZATION — The process in which substance in the solid or liquid phase is changed into the vapor phase.

VELOCITY — Speed; the rate of motion in a given direction, measured in feet per second, miles per hour, kilometers per hour, and so on.

VESSEL — A tank or container that may or may not be pressurized.

VOLATILE — A substance that readily vaporizes at a relatively low temperature.

W

WASTE LINE — A hoseline that is tied off or otherwise secured and is used to handle water in excess of that being used in a relay operation.

WATER DEPARTMENT — The municipal authority responsible for the water supply system in a community.

WATER HAMMER — A force created by the rapid deceleration or acceleration of water. It generally results from closing a valve or nozzle too quickly.

WATER SUPERINTENDENT — The manager of the water department.

WET-BARREL HYDRANT — A fire hydrant that has water all the way up to the discharge outlets. The hydrant may have separate valves for each discharge or one valve for all the discharges. This type of hydrant is only used in areas where there is no danger of freezing weather conditions.

Introduction

HISTORY OF WATER SYSTEMS

Natural water sources were a convenience as well as a living sustenance to primitive peoples. Although man soon learned to dig wells to obtain water, it was early Roman civilization that developed the first recorded means of bringing water great distances from the mountains through aqueducts (Figure I.1).

These aqueducts, which were supported on masonry arches, might well be called one of the wonders of the world. Water in the aqueducts discharged into covered masonry cisterns, and was then delivered to places of consumption primarily through pipes

Figure I.1 Aqueducts carried water in raised troughs. Sloping arches allowed the water to travel by gravity.

made of lead or, in some instances, of bored-out blocks of stone. Lead pipes measuring up to 27 inches (686 mm) in diameter have been found in Rome.

Although the Romans built aqueducts and used a volume of water equal to or in excess of the volume of water used by a city of today, the science of water supply and distribution almost vanished with the decline of the Roman Empire. A renaissance in this field did not occur until the nineteenth century.

Nothing in America could be properly called a comprehensive water system until the Philadelphia Waterworks began delivering water around the year 1800. Even the cities of New York and Boston had no waterworks worthy of the name until the middle of the nineteenth century. New York's water system consisted of an elevated iron pipe from the Harlem River down Manhattan Island to a great reservoir. From there, the water was piped along the streets through bored logs. In theory, these wooden pipes could be tapped in order to supply fire engines. This was done by removing large wooden plugs that had been inserted in the logs at various points along the route. The flow of water was usually so meager, however, that the system was seldom an effective aid to fire fighting. These plugs in the wooden pipes were known as fire plugs and this term is still commonly used to identify fire hydrants.

Reservoir supplies to assist the town water system were also beneficial for fire fighting operations. It was customary during this period to construct in the streets large cisterns, or fountains as they were sometimes called, for the use of fire engines. Fountain Street, in Providence, was so named because of the location of a fountain adjacent to it.

Early organized fire fighting operations employed "bucket brigades" to transport water to the location of a fire. The water was carried in leather buckets from the nearest water supply — usually a well, river, lake, or canal. The firefighters would then throw the water from the buckets onto the fire in an attempt to extinguish it (Figure I.2). Portable tubs or tanks with hand-operated pumps were later employed to deliver water to a fire more effectively. These water tanks were enlarged and placed on hand- or horse-drawn wheeled apparatus. Fire pumps were improved to provide double-action hand operation, but the water to the tub or tank continued to be transported from the water supply by bucket brigades. Hand-operated pumping apparatus was later equipped with a leather hose and so arranged that the pump could draft water from available supplies and pump it to the fire location. This feature eliminated the need to carry water in buckets, but it was only effective on fires that occurred relatively near a source of water. Improvements in pumps, hose, and nozzles did not greatly

Figure I.2 Early organized fire fighting operations employed bucket brigades to transport the water to the fire.

increase fire fighting efficiency except when they too were located near a source of water supply.

The installation of a system of underground water pipes extending from a central pumping station partially solved the problems of water transportation. As improvements were made in underground water systems, it was possible to make effective improvements in all fire fighting equipment. It is interesting to note that a great majority of the changes made in fire fighting equipment have been those that improved the methods of using water from distribution systems.

Today, delivering water to the scene of a fire is not only a concern in areas served by fixed water supply systems, but also in areas that are lacking such systems. Lack of a fixed water supply system requires a different approach to water supply operations. Relay pumping and tanker shuttle operations in rural areas differ greatly from urban water supply operations.

PURPOSE AND SCOPE

This manual has been prepared to provide fire service personnel with a basic understanding of waterworks systems and rural water supply operations. A study of this manual will help develop and improve understanding of the fundamental principles, requirements, and standards used to provide water for fire fighting purposes. It is also designed to give waterworks personnel an insight into the necessity of an adequate and reliable water supply for fire fighting operations.

Water Supplies for Fire Protection is not meant to serve as a complete text on the subject of water supply; rather, it deals primarily with those aspects of water supply systems and rural water supply sources and methods of concern to firefighters. This manual discusses water sources available to the fire service and types of supply systems. Information is also provided on the distribution system, which is of primary importance to the fire service. This includes requirements for size and carrying capacity of mains, hydrant specifications and maintenance procedures conducted by the fire department, and relevant maps and record-keeping procedures. Water supply calculations for fire flow, including procedures for running flow tests, and developing simple water supply curves, and available water are covered in detail. Alternative sources of water supply and delivery methods for rural and suburban fire fighting operations are also discussed in depth.

This manual prepares the firefighter to meet the water supply section requirements of the following National Fire Protection Association Standards:

- NFPA 1001, *Fire Fighter Professional Qualifications,* 1987 edition.
- NFPA 1002, *Fire Apparatus Driver/Operator Professional Qualifications,* 1988 edition.

Chapter 1

Water Supply Management

6 WATER SUPPLIES

This chapter provides information that addresses performance objectives described in NFPA 1002, *Fire Apparatus Driver/Operator Professional Qualifications* (1988), particularly those referenced in the following sections:

3-2.9

3-2.10

Chapter 1
Water Supply Management

DETERMINATION OF NEEDED FIRE FLOW

Reasons for Determination

One of the basic ingredients for sound pre-incident plans of individual buildings and high fire hazard areas is determining the needed fire flow. The required quantity of water and delivery rate will have a significant effect on attack methods, apparatus and manpower, and the need for special operations to augment inadequate supplies (Figure 1.1).

The needed fire flow is also a good measure of the adequacy of the present water system. When the needed fire flows for a community are inadequate for a large percentage of the protection area, a case can be made for critical review and potential upgrading of the water system.

Figure 1.1 Water supply systems can be used to supply large quantities of water for fire fighting operations. *Courtesy of Jay Bradish, Bradford, Pennsylvania.*

It is understandable that the needed fire flow is an important consideration. However, a point that cannot be overlooked is that regardless of the water system capability, the fire department must be able to deliver the needed fire flow through apparatus and appliances, and apply it correctly in order to provide quality fire protection.

Several methods are available to determine the needed fire flow and/or the duration that it must be available. Areas supplied by a municipal water system may utilize a method from the Uniform Fire Code Standard (Appendix A), the Insurance Services Office (ISO) *Guide for Determination of Required Fire Flow*, 1974 (Appendix B), or the needed fire flow section of ISO's *Fire Suppression Rating Schedule,* 1980 (Appendix C).

The International Fire Service Training Association does not recommend a particular method but suggests a method be selected on the basis of current fire department policy, the applicability of the standard to the fire protection area, or one that meets the requirements of the authority having jurisdiction.

PLANNING

Master Plan

An important feature of the community Master Plan is the Fire Protection Master Plan. This supplemental plan addresses the development and direction of the community's fire protection system. An integral part of the Fire Protection Plan is the concerns of present and future water supply and distribution needs. The goals to be projected for water systems should reflect the concerns listed below.

Figure 1.2 Installing fire hydrants is generally the responsibility of the Water Department.

SHORT-TERM GOALS

- Prioritizing immediate problems
- Weak system improvement
- Water main preventive maintenance program
- Old hydrant replacement and addition of needed hydrants (Figure 1.2)
- Identification of water supply resources
- Valve cycling and lubrication program
- Hydrant maintenance program (Figure 1.3)
- Identifying deficient or aging major components

LONG-TERM GOALS

- Additional sources, treatment, and pumping facilities
- Looping and gridding of distribution system
- Water system extensions and expansions
- Additional storage facilities

Figure 1.3 Routine hydrant maintenance, such as painting of hydrants, may be carried out by fire department personnel. *Courtesy of Edmond, Oklahoma Fire Department.*

- System modernization
- Sprinkler system incentive program
- Coordination of zoning and water main planning
- Future funding

Cooperation with Other Agencies

The fire department's ability to supply a wide range of emergency services on an immediate and routine basis depends on the level of cooperation and coordination maintained with other agencies. A main factor in pre-incident planning is determining the outside resources that will definitely be needed, the potential resources needed, and the manner in which they can be obtained and utilized.

When making initial contacts to utilize outside services, there are some major points that can be helpful in promoting a good relationship.

- Fire department personnel should not consider themselves as the bottom line experts and should actively seek advice in those problem areas they are attempting to resolve.

Figure 1.4 Public water utility officials can provide a great deal of assistance to the fire department.

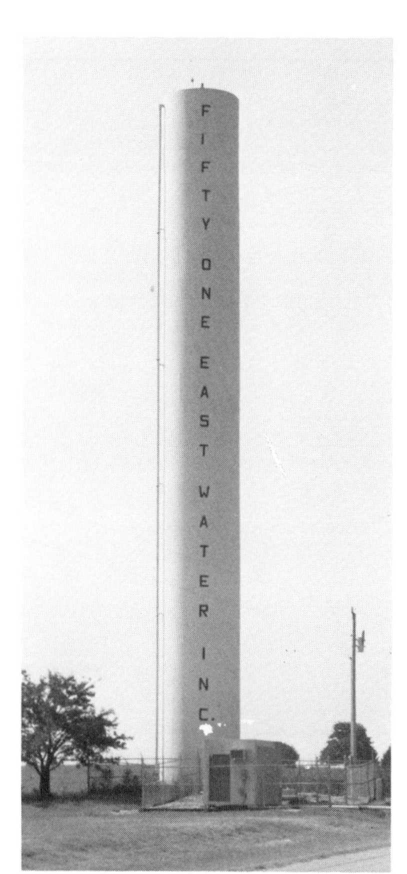

Figure 1.5 Some rural areas are served by small water supply systems. Often, these systems can supply only a very limited amount of water for fire fighting operations.

- Contacted organizations are generally very concerned with the fire protection problems within their own occupation. This gives the fire department a unique opportunity to offer reciprocal advice and support in areas beyond the immediate water supply concern.
- Interagency liaison personnel can be appointed to deal with individual contacts.
- The final and most important aspect of an interagency relationship is to maintain open and honest communication.

Some of the resources/organizations that can aid the fire department with water supply problems are

- Public water utility (Figure 1.4)
- Tank truck owners/operators
- Heavy equipment agencies and/or contractors
- Owners of private water sources (Figure 1.5)
- Military installations
- Civil Defense or emergency management/preparedness agencies
- News media organizations
  (Control of domestic usage during large-scale fire operations)
- Forest service officials
  (Airborne supply)
- Coast Guard (Figures 1.6a and b) and/or private tug boat fleet operators

The fire department's relationship with the public water utility deserves extra consideration since it represents the main bridge between the fire department and its primary extinguishing agent.

Figure 1.6a Although Coast Guard personnel are trained and equipped to protect their own ships, bases, and aircraft, they are not trained or equipped to be firefighters in the traditional sense of the term. *Courtesy of U.S. Coast Guard.*

Figure 1.6b The differences between a standard fire boat (left) and a Coast Guard vessel (right) are dramatic in terms of fire fighting capabilities. *Courtesy of U.S. Coast Guard.*

A water utility is a business, and its principal function is to provide safe water for human consumption. Meeting the requirements for reliable fire protection is usually a secondary function of the waterworks official. The objectives of the fire department can still be met under these conditions.

In smaller communities, the quantity of water used when fighting large fires is often greater than the heaviest normal water consumption; thus, it is often a problem to arrange for a sudden increase in system capacity. If fire officers are knowledgeable of what happens in a water pumping station when a fire creates a heavy water demand, they will then be more appreciative of water distribution problems. The importance of notifying the water utility when a major fire occurs is obvious. The water utility should also be informed when the fire is under control.

A water superintendent is in a position to make valuable suggestions to a fire officer when pre-incident plans are being made. The water superintendent can point out the weaker spots in the distribution system which will help to properly locate pumpers. Flow tests to determine available water for fire fighting are very important when formulating pre-incident plans. The fire officer should discuss these tests with the water superintendent because a water main extension or other change may affect the fire flow. A fire officer should make a careful survey of the area serviced by the water distribution system to determine the adequacy of fire hydrants. This survey should also determine the amount of water that may be required for any group of buildings. The heads of the two agencies should intermittently counsel with competent fire protection engineers. They may also wish to contact the ISO Commercial Risk Service, Inc. or State Fire Rating Bureau in their area. Through these conferences a long-range plan to achieve the best possible fire protection can be developed.

Maintenance should be a direct concern of both agencies. The fire department should report to the water utility, in writing, all components that are not working properly. Likewise, it is crucial that the water utility notify the fire department of maintenance that requires water mains or hydrants to be out of service, as well as how long the service will take, and when restoration is complete.

When a water utility is privately owned, the utility may have a franchise or contract with a city or the state. These franchises give a water utility certain basic rights along with a responsibility to provide proper service to customers for personal and industrial uses. They may also be required to provide for adequate fire protection. The fire chief or water supply officer, through proper channels, should initiate negotiations with the water authority on matters of importance to the fire department.

Another area of planning is to maintain contact with local construction contractors. Development areas will require a review of construction types and installations to determine potential water requirements beyond the basic fire flows. Fire departments should realize this may be a touchy area and will require diplomacy to obtain the necessary information (Figure 1.7).

The fire department may want to consider contacting the local branch of the U.S. Department of Agriculture, Soil Conservation Servicc. The Soil Conservation Service in some areas provides technical assistance to the installers of static water sources in rural areas. The Soil Conservation Service may require permit approval from the respective state water organization to obtain matching funds for the construction of larger sources. They may also have requirements regarding the location of or access to rural static sources. A working relationship between the fire de-

Figure 1.7 Personnel trained in proper plans review procedures are a valuable asset to fire protection planning.

partment and the Soil Conservation Service could promote a better understanding of the requirements for access and location that makes a rural water source workable for the fire department.

Pre-Incident Planning

The backbone for a quality planning structure is a system of prefire or pre-incident plans. The water supply considerations of a pre-incident plan will have a direct effect on the other components of the plan. Apparatus, suppression strategies, and manpower are some of the areas influenced by the water supply capability. A review of a given occupancy will require the following water supply information:

- Required fire flow
- Available fire flow
- Location(s) of supply
- Reliability of supply
- Auxiliary supply
- Water supply utilization methods

A good pre-incident plan is one of the most beneficial tools in a strong first defense of an occupancy or incident. When using a pre-incident plan as a fireground guide, consideration should always be given to the following factors:

- Change in occupancy
- Change in building contents

- Additional exposures (Figure 1.8)
- Water system repairs
- Weather conditions that deter the use of the pre-incident plan (Figure 1.9)

Figure 1.8 Pre-incident plans should include methods for protecting adjoining or adjacent exposures.

Figure 1.9 Locating hydrants may be difficult after heavy snowfalls.

FIRE DEPARTMENT WATER SUPPLY OFFICER

All fire departments, regardless of size, should consider designating a water supply officer. This individual would be responsible for overseeing the water supply needs of the department. The water supply officer may be assigned to the position on a full-time basis or as a staff position with special water supply duties in addition to other duties. The water supply officer should be knowledgeable in the following areas:

- Public water supplies and distribution systems
- Static source utilization
- Hydraulics and fire flow calculations
- Water maps and plans
- Fire fighting operations
- Planning skills
- Training techniques
- Communication skills

The duties of a water supply officer are many and varied. A major function is to act as liaison between the fire department and water department. The administrative responsibilities include the following:

- Prepare reports identifying the fire department's water needs as determined by an evaluation of the required fire flow.

- Work with water utility representative on plans for extension of distribution system.
- Supervise the inspection and testing of fire hydrants, private connections, and other areas.
  — Reports of such inspections and tests should be made to the water utility representative, and the water supply officer should see that maintenance work is done.
- Assist with the development of the water supply portion of pre-incident plans.
  — Information regarding the water supply in a given area should be made available to the officers involved with planning in their district.
- Inform and train fire officers concerning the water supplies in their district.
- See that detailed maps and records are kept of all potential static water sources.
  — Maps should show source location, capacity, accessibility and special requirements (suction hose, adapting accessories). A card file or computer stored information should give clear directions to source and contain all pertinent information including potential problems regarding source (Figure 1.10).

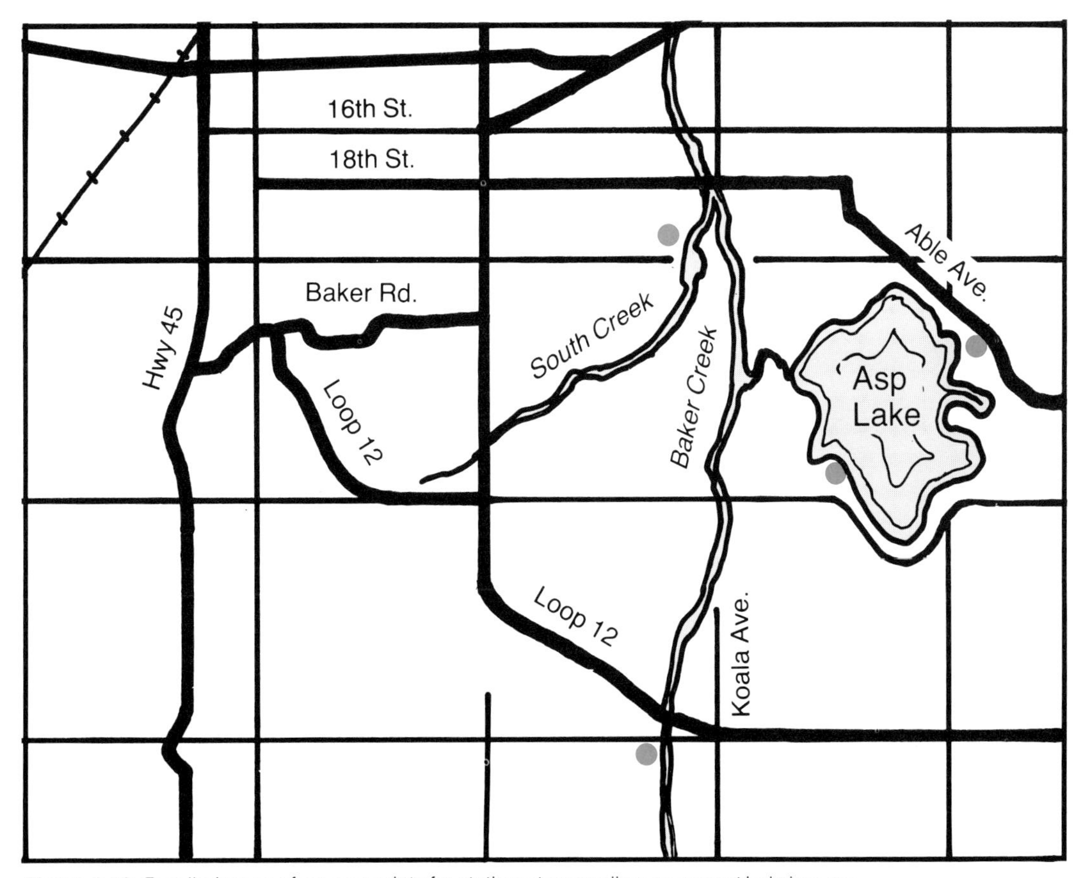

Figure 1.10 Detailed maps of access points for static water supplies are a great help in areas not protected by a water system.

- See that a maintenance record of static sources is kept and updated.
  - A seasonal check of sources will help assure that they are usable and will define new problems (silt, blocked access, not deep enough for draft, dried up). All sites should be numbered and marked with suitable signs.
- Prepare fireground standard operating procedures (SOP's) for water relay or tanker shuttle operations.
  - These procedures should include policies on locating pumpers at the fire and water source. SOP's should also be developed for rapid filling, unloading of tankers, and instructions for drivers in areas having road problems.

Special operating plans must be prepared for locations known to have a water supply deficiency and for community-wide fire protection service should a major water system failure occur. In some departments, the water supply officer has the responsibility of reviewing new building plans for adequate sprinkler or standpipe supply and connections. Private water systems may also have to be reviewed. The water supply officer may be responsible for overseeing or performing the testing and inspection of new systems before they are placed into operation.

Other administrative functions of the water supply officer occur in areas with an inadequate or no distribution system where the fire department must rely on static sources.

An area that may pose some difficulty for fire departments is the legality of obtaining private water for fire fighting operations. The water supply officer must check to determine what powers the fire department has to use private water. Written agreements, refill provisions, or similar arrangements may be necessary, especially where agriculture is dependent on private supplies.

Routing of apparatus is a consideration of the water supply officer. The water supply officer must consider the maximum height, width, power capabilities, and weight of the department's vehicles when deciding areas of access. Certain roads, hills, bridges, or terrain may have to be designated "off limits" for specific pieces of apparatus. Urban pumpers are not designed for cross-country terrain and may be severely damaged if taken off road. Winding roads must also be considered due to the characteristics of tanker handling. Sharp turns that would cause sudden water shifts should be avoided.

The water supply officer should see that access points are available to as many static sources as possible or economically feasible. Paved or gravel ramps, dry hydrants, or other means of access to points of high water use will greatly simplify drafting operations. The water supply officer should see that all installations are proper and that they meet with the owner's approval if done on private property.

Fireground Responsibilities

The fireground functions of a water supply officer are entirely different from the administrative areas. It is desirable to have the same individual performing both functions; however, the fireground operation may be handled by a second-in company officer, a designated chief officer, or possibly a mutual aid officer. The fireground functions of a water supply officer working with a water system are

- Report to the fireground commander (FGC) and assume control over all water supply operations.
- Determine from the fireground commander where water is needed.
- Determine from where and how water will be supplied.
- Supervise placement and layout of apparatus and supply lines. (Consideration must be given to water system layout and location of sprinkler, standpipe, and private supplies.)
- Keep a working knowledge of the amount of water remaining available, and notify fireground commander when it approaches critical.
- Oversee pump operations and make adjustments to maximize pump operations and return unneeded companies to service. (Arrange for gas, diesel, oil, and manpower relief.)
- Work with water department to clear or minimize water supply problems at the scene.
- Plan for contingencies.
- Arrange for additional water supply apparatus and supplemental water supply operations as needed (Figure 1.11).

Figure 1.11 Staging areas are extremely important when there are a large number of incoming apparatus waiting to be assigned. *Courtesy of Ron Coleman.*

- Prepare for mechanical breakdown or burst hose. (Maintain contact with chief engineer or mechanic on scene to foresee mechanical problem.)
- Maintain contact with pump operators to ascertain problems and maintain morale (Figure 1.12).
- Inform fireground commander when all resources have been committed.
- Record placement and layout for post-fire critique and suggest appropriate changes if necessary.

Figure 1.12 The officer in charge of water supply operations should contact relay driver/operators frequently to ensure that everything is in order.

The fire department may also wish to consider establishing a water supply company (WSC). The water supply company is an engine company and crew designated to assist the water supply officer. Some of the functions they perform may include scanning the water system maps or area to locate available hydrants, assist in positioning incoming apparatus, and in general, being available for assignment or utilization by the water supply officer.

The requirements for a nonsystem water supply area are the same as those in a system supplied area with a few major exceptions. There will be no water utility personnel to maximize the system's potential. Related to this, and because of the generally increased relay lengths and/or the distance covered in tanker shuttle operations, a second level of command must be established under the water supply officer. Within a long relay operation, additional officers must be appointed to cover segments of the relay and report conditions to the water supply officer. In tanker shuttle operations, specific officers will be appointed with an operating crew under their command as shown in Table 1.1.

TABLE 1.1
SHUTTLE OPERATION RESPONSIBILITIES

FILL SITE OFFICER	UNLOADING SITE OFFICER
Traffic Control	Traffic Control
Fill Site Control	Unloading Site Control
Communications	Communications
Preparation	Preparation
Positioning	Positioning
Personnel	Personnel
Vent Operator	Vent Operator
Pumper/Hydrant Operator	Valve or Pump Operator
Make and Break Personnel	Make and Break Personnel
Overall Safety	Overall Safety

The fill site officer and unloading officer perform similar functions at their respective end of a shuttle. The responsibilities include the following:

- Provide Traffic Control Personnel
  - To maintain traffic flow of emergency vehicles and to regulate and reroute civilian traffic.
  - To maintain an orderly flow in and out of the fill or unloading area.
  - To provide back-up personnel or self for backing operations.

- Fill or Unloading Site Control
  - Communications. Convey instructions to drivers approaching the site, to draft pumper operators (or unloading pumper operators if used), and relay conditions to the water supply officer.
  - Preparation. Determine site locations and supervise setup procedures and equipment.
  - Positioning. Includes positioning of draft pumpers, fill or unloading lines, portable tanks, and establishing a staging area for rotating tankers to hold at until the site officer calls for them.

- Personnel for Pumper/Hydrant Operator
  - The draft pumper operator or the pump operator at hydrant connected pumper (possibly hydrant or valve operator when directly filling tankers from hydrant).
  - Unload Operator. Unload site pump operator or person controlling tanker dump valves.
  - Make and Break Personnel. Individuals at fill or unload site who make pumper or tanker connections and manipulate hoselines as needed.
  - Vent Operator. Person charged with opening vents on filling or unloading tankers. Must also make sure vent remains open during filling or unloading.

Safety

Safety in all evolutions is paramount. Safety should be constantly stressed in both the classroom and during hands-on training. In water supply evolutions, areas to be observed include the following:

- Use of proper adapters, tools, and fittings
- Laying or handling hose and couplings
- Personnel in full protective gear (Figure 1.13)

Figure 1.13 Firefighters should always wear full protective equipment and carry any tools they might commonly need when they enter a fire building.

- Proper lifting and pulling techniques (moving hose or portable appliances) (Figure 1.14)
- Positioning of apparatus and portable tanks
- Fill or unloading hoselines under pressure (properly secured or controlled)

Figure 1.14 Firefighters should use proper lifting techniques whenever working with heavy objects, such as this hard suction hose. Remember to bend with the knees, not the back.

- Hose appliances (clamps, valves, etc.)
- Driving of apparatus
- Working on or about apparatus
- Mounting and dismounting
- Operation of top-mounted vents and fills (Figure 1.15)

The water supply officer should remain near the fire scene during tanker shuttle operations so an ongoing evaluation of the needed suppression water supply can be conducted. Because of this fact, the water supply officer may also serve as the unloading site officer in situations where there is a limited amount of manpower or available officers.

Figure 1.15 In addition to fixed fill piping, tankers may also be filled through the tank vents on the top of the vehicle.

TRAINING

The ability to provide adequate water supplies on the fireground hinges on proper training. Drills must be conducted with all the apparatus, new techniques mastered, and most importantly, procedures learned so they become a natural sequence at a fire.

Location

The location of a training session affects the material covered and the method in which it is presented. When dealing with water supply, the areas covered will depend on whether the setting is in a classroom or hands-on.

- Classroom (Figure 1.16)
  - — Water movement theory, hydraulics
  - — Pump operations theory (relay, hydrant, and draft operation)
  - — Explanation of relay and shuttle procedures
  - — Review of water source locations
  - — Officer's duties and responsibilities
  - — Safety

Figure 1.16 Theoretical aspects of water supply operations may be dealt with in the classroom, prior to "hands-on" training.

- Hands-On (Figure 1.17)
  - — Actual relay and shuttle operation
  - — Practical pump operations
  - — Filling and unloading operations
  - — Mutual aid evolutions
  - — Measure of actual vs. flows from relays and shuttles

Figure 1.17 It is important to practice water shuttle operations regularly if they are to work properly.

Frequency

The frequency of training will be a decisive factor in the fireground capabilities of any department. This holds especially true in dealing with water supply operation. A fire department may have several multiple alarm fires every year while another has a major fire only once in five years. In both situations, the amount of training should be sufficient to provide the best water supply possible.

Standard Operating Procedures

One of the most important aspects of training is the use and familiarization with standard operating procedures (SOP's). Refer to Appendix D for suggested SOP's.

The fire department should practice water supply operations that are up to their maximum capability. When possible, this should include mutal aid response. The procedure should be to train for the largest fire expected and then one a little bigger. By utilizing this technique, several advantages are realized

- Best potential for providing necessary fireground flow
- Less likely to underestimate fire potential or necessary water supply
- Rapid knockdown will be achieved on medium and small fires
- Officer judgment will prevent overkill in small fire situations

Mutual Aid

Fire departments that have agreements with neighboring departments for mutual aid should perform evolutions with these

Figure 1.18 In areas where water shuttle operations commonly involve different fire departments, mutual aid training exercises are vital for efficient and effective operations.

departments (Figure 1.18). Interdepartmental training will help assure that procedures flow more smoothly on combined water supply operations. Areas of importance include the following:

- Variance in SOP's
- Preparation for equipment differences (hose threads, tanker and pumper connections, adapters)
- Jurisdictional command procedures
- Communications
- Primary water source locations
- Intercompany familiarization

Communications

Mutual aid responses require fire departments to communicate water supply information at varying points during an incident (fill site location, tanker routes, apparatus positioning for filling and dumping, how to fill or empty apparatus). Training for fireground communications should include the following:

- Standardized frequency, terminology
- Apparatus numbering and recognition
- Command and/or water supply frequencies
- Mutual aid call procedures for dispatchers

Driver/Operators

The importance of defensive driving practices during water supply operations (especially tanker shuttles) cannot be overemphasized. Tanker apparatus require special consideration due to their weight, higher center of gravity, size, and potential for large load shifts. These factors cause a variety of handling differences compared to other types of apparatus.

- Longer stopping and braking distances
- Poor cornering abilities
- Increased difficulties in backing and positioning
- Less stability on poor roads
- Increased potential for rollover
- Less acceleration
- Added problems in inclement weather

Driver/operators of tankers should be given extensive training and nonemergency driving time to become familiar with the characteristics of the heavier apparatus. Training drills and procedures for drivers are listed in IFSTA, **Introduction to Fire Apparatus Practices,** and IFSTA, **Pumping Apparatus Practices.**

Safe driving practices should also extend to shuttle training evolutions. Shorter distances, familiar roads, reduced traffic, and good weather are not justification for reckless driving and require the same caution given any emergency response.

Chapter 1 Review

Answers on page 259

TRUE-FALSE: Mark each statement true or false. If false, explain why.

1. The water utility's primary purpose is to provide adequate water supply for fire protection.

   ☐ T ☐ F ____________________

   ____

2. The purpose of a fire protection master plan is to address the development and direction of the fire department.

   ☐ T ☐ F ____________________

   ____

3. The fire department and the municipal water department are the only two resources available to correct water supply problems or deficiencies.

   ☐ T ☐ F ____________________

   ____

4. Administrative responsibilities of the water supply officer include preparing standard operating procedures for water supplies, training fire officers regarding water supplies in their districts, and supervising inspection and testing of hydrants.

   ☐ T ☐ F ____________________

   ____

5. The fireground commander is responsible for traffic control in the area of a water shuttle fill site.

   ☐ T ☐ F ____________________

   ____

MULTIPLE CHOICE: Circle the best answer.

6. Maintenance is a direct concern of __________.
   A. fire department
   B. water department
   C. both

7. Which of the following is *not* a short-term goal of a fire protection master plan with regard to water supply?
   A. Water main preventive maintenance program
   B. Valve cycling and lubrication program
   C. Replacement of old hose on water supply engine companies
   D. Hydrant maintenance program

8. The two primary functions of a water utility are to provide safe water for human consumption and to ________.

A. provide adequate water for industrial processes
B. meet the requirements for reliable fire protection
C. prevent water system equipment failure
D. make water system repairs

9. Which information about water supplies is *not* required when preparing a pre-incident plan?

A. The required fire flow
B. Location of supply
C. Auxiliary supply
D. Number of pumps normally operating at the treatment plant

10. Which one of the following *is* a fireground responsibility of the fire department water supply officer?

A. Determine from fireground commander where water is needed
B. Size of attack lines to be used
C. Placement of master stream devices
D. Calculation of potential damage to building due to weight of water used for extinguishment

11. Which one of the following duties is *not* the responsibility of the unloading site officer?

A. Supervision of unloading site make and break personnel
B. Optimum positioning of the fill pumper
C. Communication with apparatus approaching the fill site
D. Positioning of the portable water tank

12. Which of the following is *not* an important mutual aid consideration when operating water shuttles?

A. Standardized models of radios
B. Standardized radio frequency and terminology
C. Command and/or water supply frequencies
D. Mutual aid call procedures for dispatchers

SHORT ANSWER: Answer each item briefly.

13. Name the three guides for determining needed fire flow utilizing a municipal water supply system.

A. ____________________

B. ____________________

C. ____________________

14. Name the water supply standard that applies to areas without a municipal water supply system.

____

____

15. Name at least three long-term goals pertaining to water supply that should be included in a fire protection master plan.

______

______

______

16. What are three ways the fire department can help promote good relationships with other organizations involved in water supplies?

______

______

______

LISTING

17. List three factors that must be considered when using a pre-incident plan as a fireground guide.

______

______

______

18. List five safety concerns that should be included in tanker shuttle training.

______

______

______

______

______

19. List three areas that need to be addressed when different departments participate in tanker shuttle training.

______

______

______

Chapter
2

Water System Fundamentals

This chapter provides information that addresses performance objectives in NFPA 1001, *Fire Fighter Professional Qualifications* (1987) and NFPA 1002, *Fire Apparatus Driver/Operator Professional Qualifications* (1988), particularly those referenced in the following sections:

NFPA 1001

Fire Fighter II

4-15.1

4-15.2

4-15.4

4-15.5

Fire Fighter III

5-15.1

5-15.5

5-15.6

NFPA 1002

3-2.1

3-2.4

3-2.6

3-2.7

Chapter 2
Water System Fundamentals

Since water is the primary agent used for fire extinguishment, it is critical that ample supplies be delivered to the fire scene. To do this, water departments use a variety of complex delivery systems. An understanding of the theory of pressure, how it is developed, and how it affects water movement is necessary to comprehend these systems. Well-designed systems have features to create pressure to move the water from the source throughout the distribution mechanism for fire fighting. A description of the components of the distribution system completes the firefighter's knowledge of these systems. This fundamental information will begin with the types of pressure, how they act on fluids, and how pressure can be developed.

HYDRAULICS

Theory of Pressure

The word *pressure* has a variety of meanings. Ordinarily one thinks of pressure as an application of force on one thing by another thing. Pressure further suggests such synonyms as compression, pushing, or squeezing, and has often been used to express a burden of physical or mental stress.

Pressure can easily be confused with *force*. Force is the simple measure of weight, usually expressed in pounds or kilograms. Pressure is *force per unit area*. The measurement of force by weight is directly related to the force of gravity, which is the amount of attraction the earth has for all bodies. If a given number of objects of the same size and weight are placed upon a flat surface, they will each exert an equal force on that surface. For example, three square containers of equal size (1 x 1 x 1 foot [0.3 m by 0.3 m by 0.3 m]) containing 1 cubic foot of water (0.1 m^3) and weighing 62.5 pounds (28 kg) each are placed as illustrated in Figure 2.1. Each container exerts a force of about 62.5 pounds per

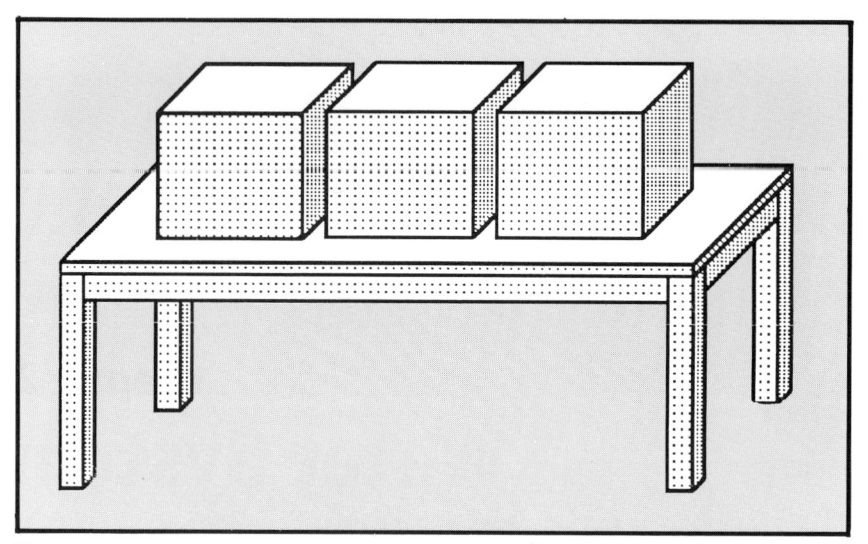

Figure 2.1 Each container measures 1 cubic foot (0.1 m^3) and weighs 62.5 pounds (28 kg). Each container therefore exerts a pressure of 62.5 pounds (28 kg) per square foot (0.1 m^2).

5-90 square foot (psf) (3 kPa) with a total of about 187.5 pounds (85 kg) of force over a 3 square foot (0.3 m^2) area. Pressure may be expressed in pounds per square foot (psf [Pa]) or pounds per square inch (psi [kPa]. If the containers are stacked on top of each other as shown in Figure 2.2, the total force (187.5 pounds [85 kg] will remain the same, but the area of contact is reduced to 1 square foot (0.1 m^2). The pressure then becomes 187.5 pounds per square foot (9 kPa). 5-90

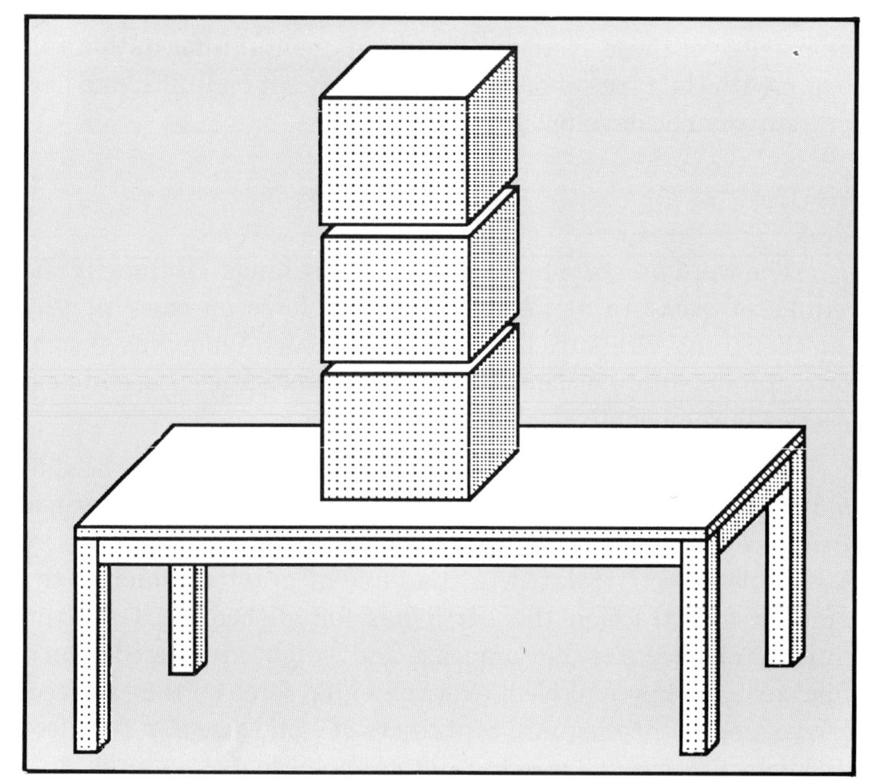

Figure 2.2 Stacked containers exert more pressure than they do singly. Each of these containers measures 1 cubic foot (0.1 m^3) and weighs 62.5 pounds (28 kg). By being stacked, they exert a pressure of 187.5 pounds (85 kg) per square foot (0.1 m^2).

5-90

The weight of a cubic foot (0.028 m^2) of water is approximately 62.5 pounds (28 kg). Since 1 square foot (0.1 m^2) contains 144 square inches (92 903 mm^2), the weight of water in 1 square inch (645 mm^2) 1 foot (0.3 m) high equals 62.5 divided by 144, or 0.434 pounds (0.2 kg). A 1-square-inch (65 mm^2) column of water 1 foot (0.3 m) high will therefore exert a pressure at its base of 0.434 psi (3 kPa). The height required for a 1-square-inch (65 mm^2) column of water to produce 1 psi (7 kPa) at its base equals 1 divided by 0.434, or 2.304 feet (0.7 m); therefore, 2.304 feet (0.7 m) of water will exert a pressure of 1 psi (7 kPa) at its base, as shown in Figure 2.3.

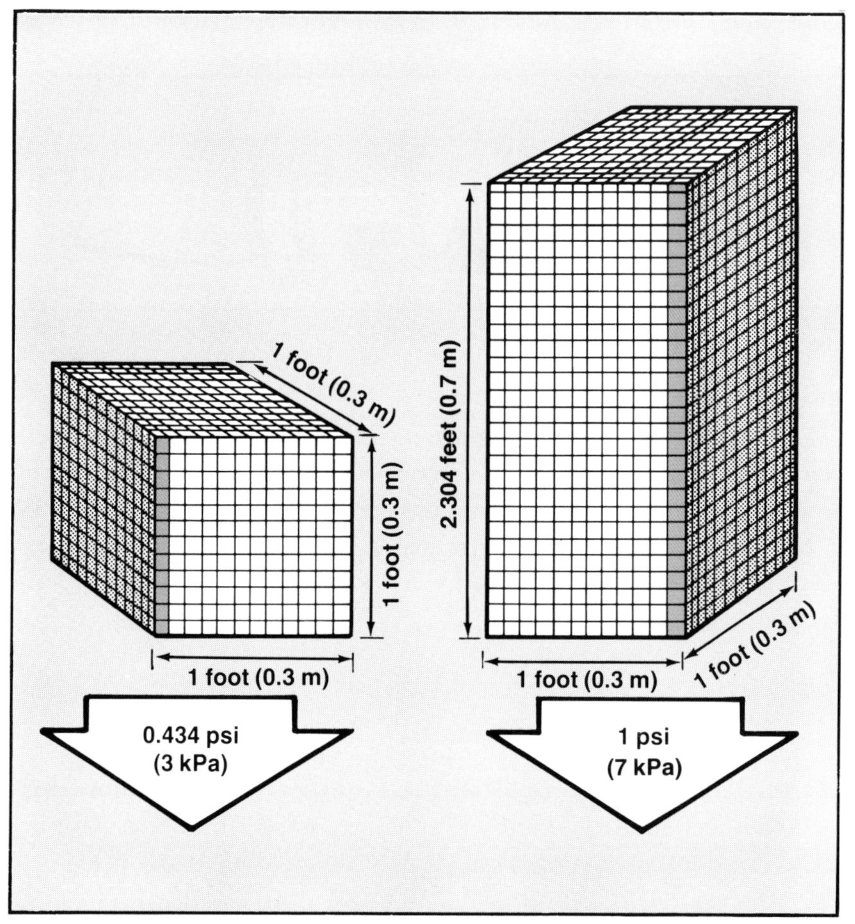

Figure 2.3 One cubic foot (0.1 m^3) of water contains 1,728 cubic inches (283 167 m^3) of water. The pressure exerted by a column of water 1 foot (0.3 m) high is 0.434 psi (3 kPa). To exert a pressure of 1 psi (7 kPa), a column of water must be 2.304 feet (0.7 m) or 27½ inches (699 mm) high.

HOW PRESSURES ACT ON FLUIDS

The speed with which a fluid travels through hose is developed by pressure upon that fluid. The speed of travel is often referred to as *velocity*, and the kind of pressure should be identified because the word *pressure* in connection with fluids has a very broad meaning. There are six basic principles that determine the action of pressure upon fluids, and they should be clearly understood before the kinds of pressures are studied.

Figure 2.4 Fluid pressure is perpendicular to any surface on which it acts.

The First Principle

Fluid pressure is perpendicular to any surface on which it acts. This principle is illustrated by a vessel having flat sides and containing water. The pressure exerted by the weight of the water is perpendicular to the walls of the container. If this pressure is exerted in any other direction, as indicated by the slanting arrows, the water would start moving downward along the sides and rising in the center (Figure 2.4).

The Second Principle

Fluid pressure at a point in a fluid at rest is of the same intensity in all directions. It also can be stated that fluid pressure at a point in a fluid at rest has no direction (Figure 2.5).

Figure 2.5. In a fluid at rest, fluid pressure is of the same intensity in all directions.

The Third Principle

Pressure applied to a confined fluid from without is transmitted equally in all directions. This principle is illustrated by a hollow sphere with a water pump attached. A series of gauges is set into the sphere around its circumference. With the sphere filled with water and pressure applied by the pump, all gauges register the same (Figure 2.6).

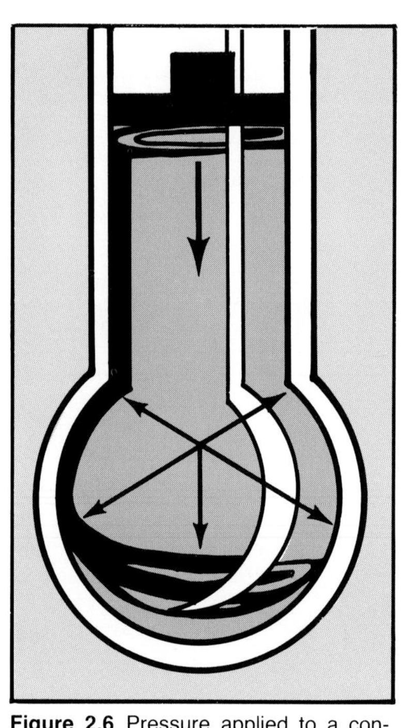

Figure 2.6 Pressure applied to a confined fluid from without is transmitted equally in all directions.

The Fourth Principle

The pressure of a liquid in an open vessel is proportional to its depth. This principle is illustrated by three vertical containers each being 1 square inch (65 mm^2) in cross-sectional area (Figure 2.7). The depth of the water is 1 foot (0.3 m) in the first container, 2 feet (0.6 m) in the second, and 3 feet (0.9 m) in the third. The pressure at the bottom of the second container is twice that of the first and the pressure at the bottom of the third container is three times that of the first. Thus, the pressure of a liquid in an open container is proportional to its depth.

The Fifth Principle

The pressure of a liquid in an open vessel is proportional to the density of the liquid. This principle is illustrated by two containers, one holding mercury 1 inch (25 mm) in depth, the other holding water 13.55 inches (344 mm) in depth (Figure 2.8). The pressure at the bottom of each container is approximately the same since mercury is approximately 13.55 times denser than water. Hence,

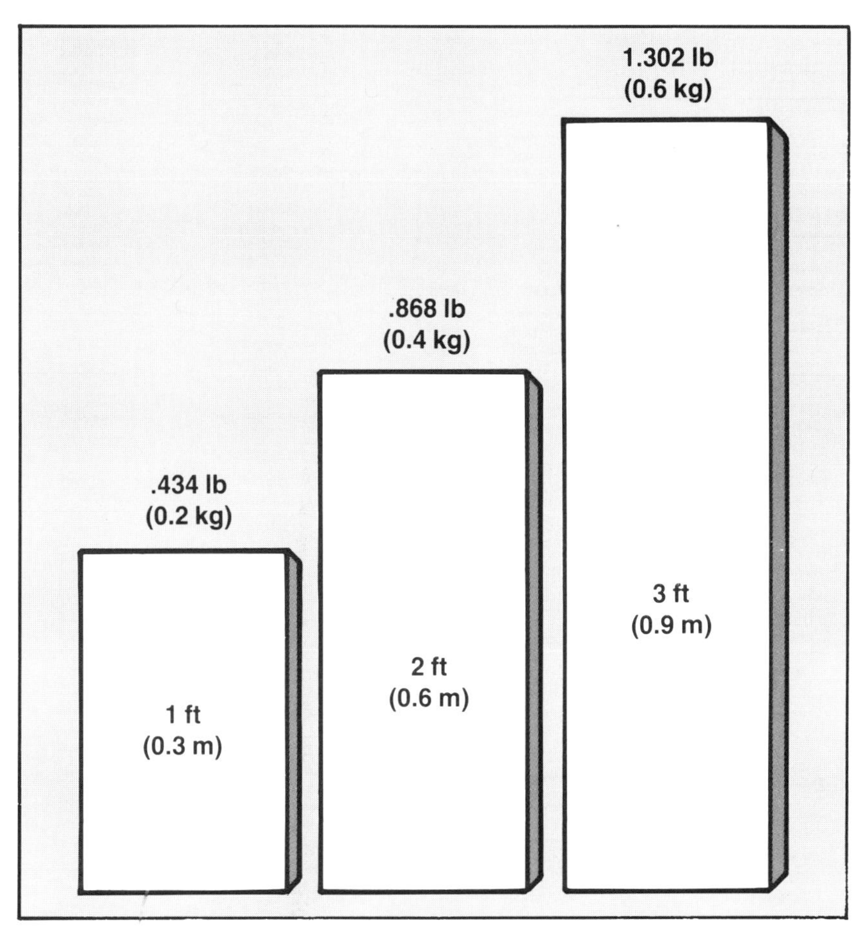

Figure 2.7 The pressure of a liquid in an open vessel is proportional to its depth.

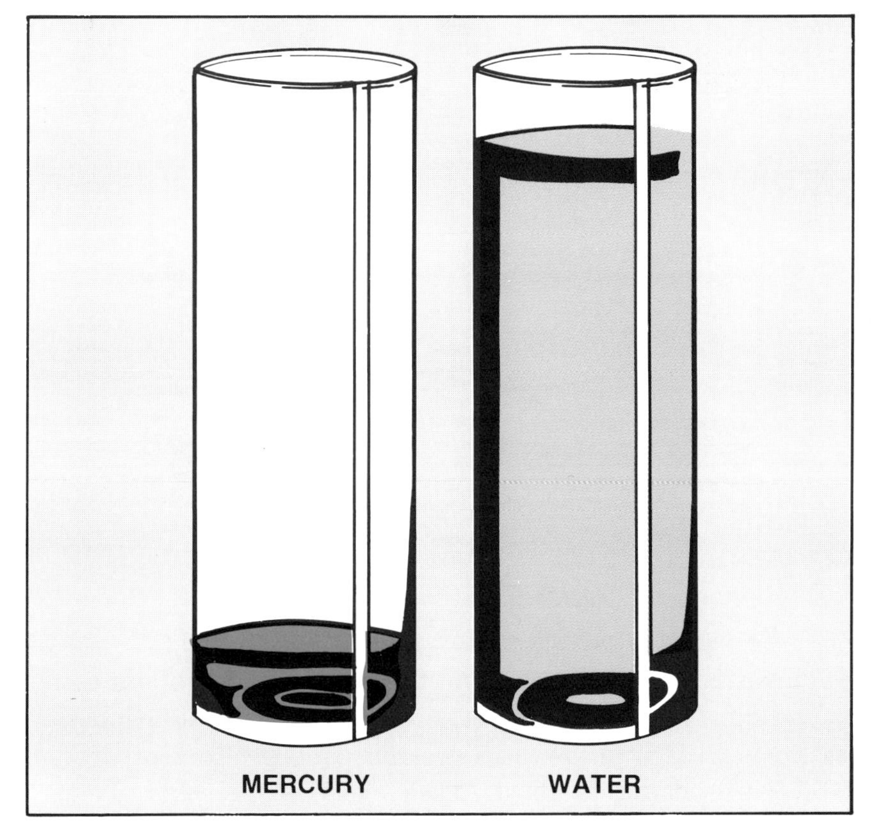

Figure 2.8 The pressure of a liquid in an open vessel is proportional to the density of the liquid.

the pressure of a liquid in an open vessel is proportional to the density of the liquid.

The Sixth Principle

The pressure of a liquid on the bottom of a vessel is independent of the shape of the vessel. The principle is illustrated by a number of containers of various shapes, each having the same cross-sectional area at the bottom and having the same height (Figure 2.9).

The pressure exerted at the bottom of each will be exactly the same if each vessel is filled with the same kind of liquid. The pressure of the liquid on the bottom of a vessel is, therefore, independent of the shape of the vessel.

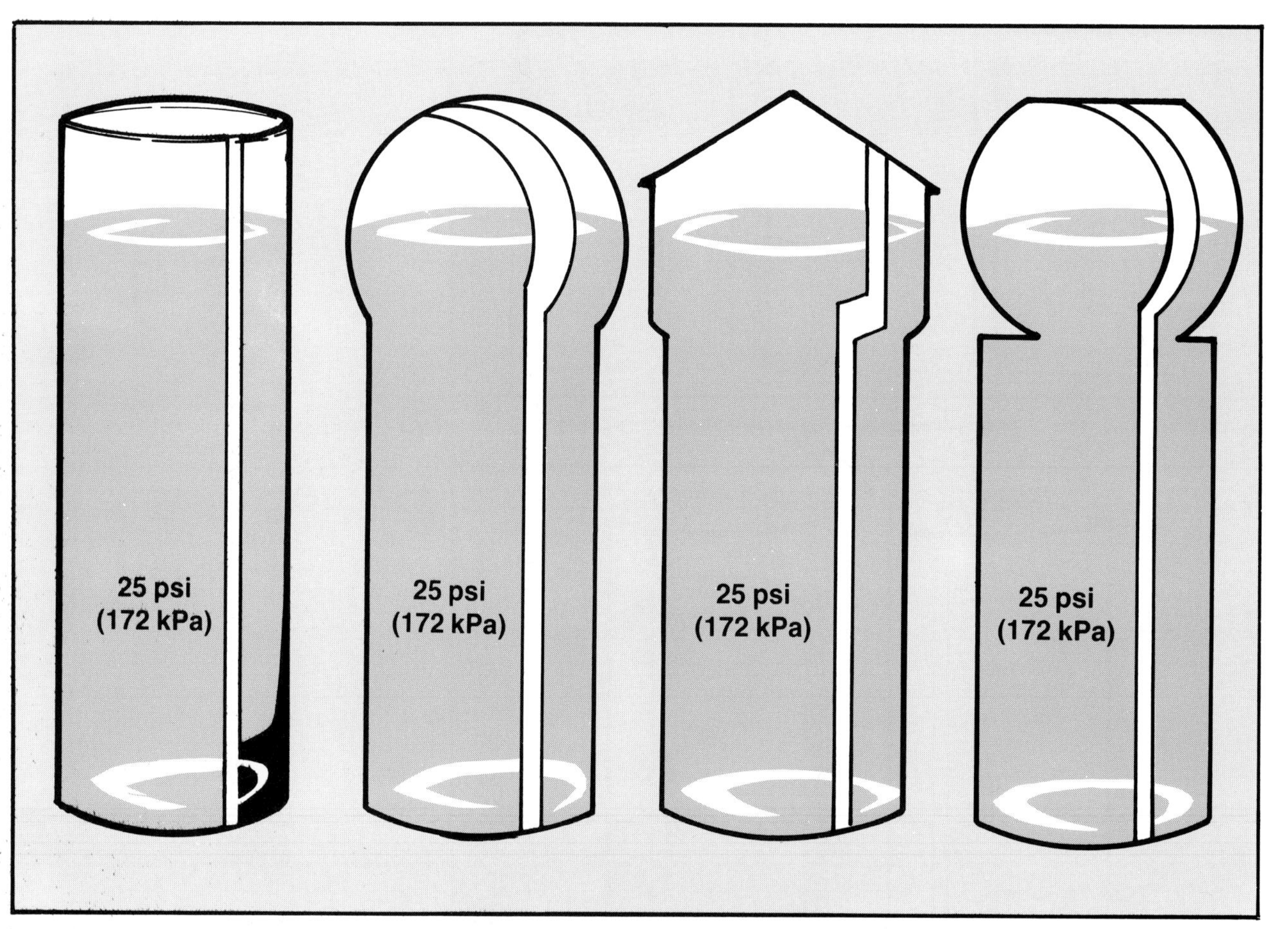

Figure 2.9 The pressure of a liquid on the bottom of a vessel is independent of the shape of the vessel.

TYPES OF PRESSURE ON FLUIDS

Atmospheric Pressure

The atmosphere surrounding the earth has depth and density and exerts pressure upon everything on earth. Atmospheric pressure is greatest at low altitudes; consequently, its pressure at sea level is used as a standard. At sea level the atmosphere exerts a pressure of 14.7 psi (101 kPa).

A common method of measuring atmospheric pressure is by comparing the weight of the atmosphere with the weight of a column of mercury: the greater the atmospheric pressure, the taller the column of mercury. The pressure of 1 psi (6.9 kPa) will make the column of mercury about 2.04 inches (51.8 mm) high. Therefore at sea level, the column of mercury will be 2.04 x 14.7, or 29.9 inches (759 mm) (see Figure 2.10).

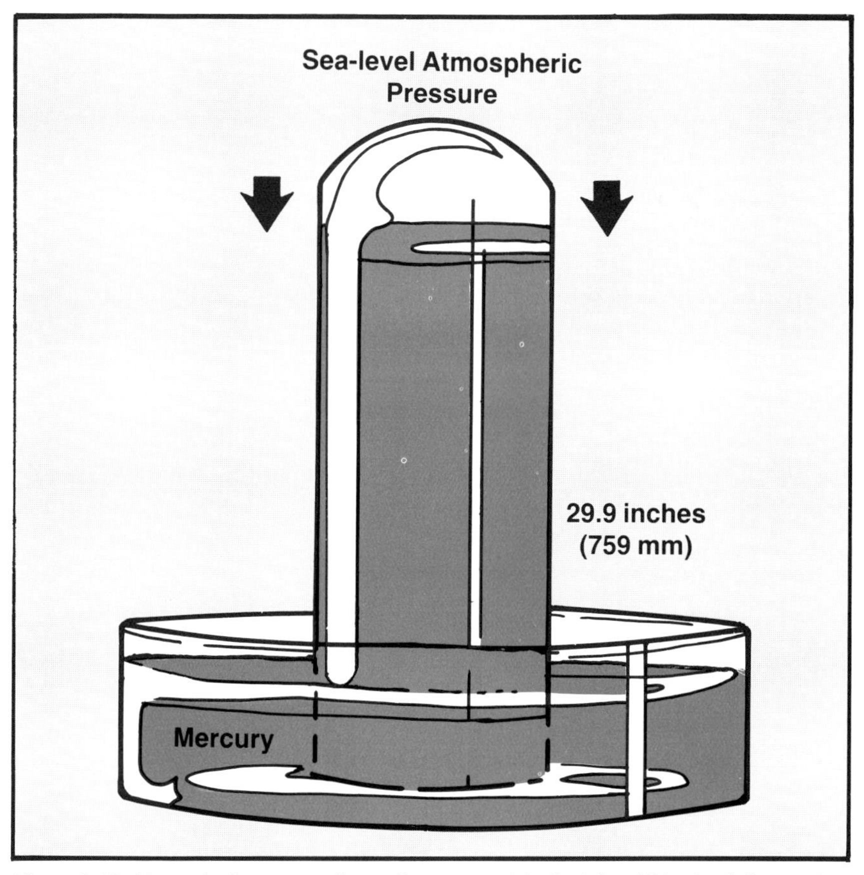

Figure 2.10 Atmospheric pressure forces the mercury into the tube of this simple barometer. A pressure of 1 psi (7 kPa) causes the mercury column to rise 2.04 inches (52 mm).

Most pressure gauges read pounds per square inch with the atmospheric pressure considered but not included. For example, a gauge reading 10 psi (69 kPa) at sea level is actually indicating 24.7 psi (170 kPa) (14.7 + 10). Engineers make the distinction between the actual atmospheric pressure and a gauge reading, 10 psi (69 kPa), by writing *psig (kPag)*, which means pounds per square inch gauge. The notation for actual atmospheric pressure, 24.7, is *psia, (kPaa)* which means pounds per square inch absolute (the psi [kPa] above a perfect vacuum, absolute zero). Throughout this manual *psi (kPa)* means *psig (kPag)*.

When a gauge reads negative 5 psig (34 kPa) it is actually reading 5 psi (34 kPa) less than the existing atmospheric pressure (at sea level, 14.7 -5, or 9.7 psia [67 kPa]). Any pressure less than atmospheric pressure is called vacuum. Absolute zero pressure is called a *perfect vacuum*.

Head Pressure

Head, in the fire service, refers to the height of a water supply above the discharge orifice. In Figure 2.11 the head is 100 — the top of the water supply is 100 feet (30 m) above the hydrant discharge opening. To convert head in feet to head pressure, all that has to be done is to divide the number of feet by 2.304 (the number of feet that 1 psi (7 kPa) will raise a column of water). The water source in Figure 2.11 has a head pressure of 43.4 psi (299 kPa) (Table 2.1).

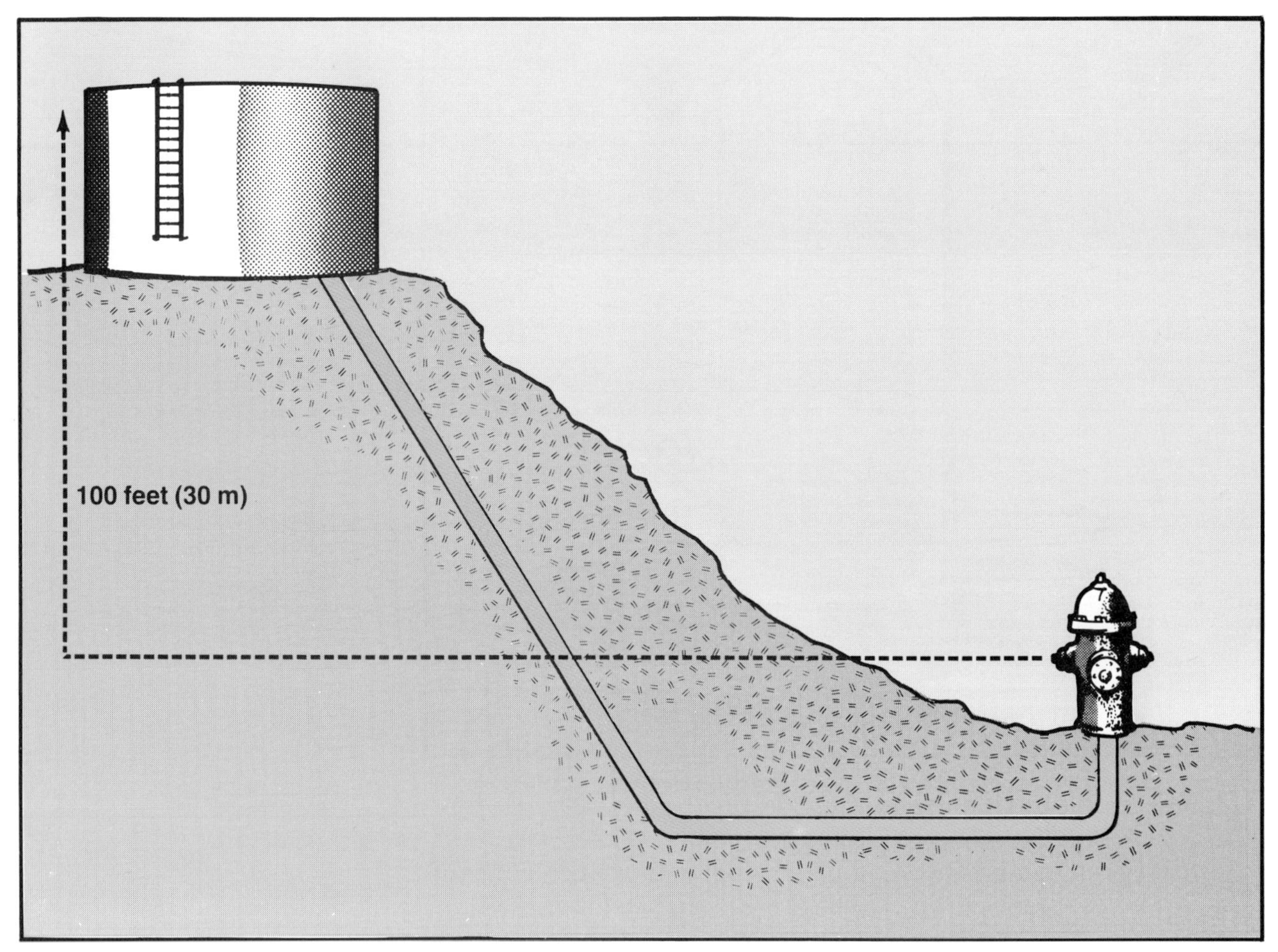

Figure 2.11 The head in this illustration is 100 feet (30 m). The head pressure is 43.4 psi (299 kPa).

Static Pressure

The word "static" means at rest or without motion. This pressure may be produced by an elevated water supply, by atmospheric pressure, or by pumps applied to a closed system in which no water is flowing. A water flow definition of static pressure is "STATIC PRESSURE IS STORED OR POTENTIAL ENERGY THAT IS AVAILABLE TO FORCE WATER THROUGH PIPE AND FITTINGS, FIRE HOSE, AND ADAPTERS." A true condition of static pressure is seldom found in water distribution systems since water for domestic and/or industrial supply is almost constantly being used.

TABLE 2.1
HEAD IN FEET AND HEAD PRESSURE

Feet of Head	Pounds per Square Inch	Pounds per Square Inch	Feet of Head
5	2.17	5	11.50
10	4.33	10	23.00
15	6.50	15	34.60
20	8.66	20	46.20
25	10.83	25	57.70
30	12.99	30	69.30
35	15.16	35	80.80
40	17.32	40	92.30
50	21.65	50	115.40
60	26.09	60	138.50
70	30.30	70	161.60
80	34.60	80	184.70
90	39.00	90	207.80
100	43.30	100	230.90
120	52.00	120	277.00
140	60.60	140	323.20
160	69.20	160	369.40
200	86.60	180	415.60
300	129.90	200	461.70
400	173.20	250	577.20
500	216.50	275	643.00
600	259.80	300	692.70
800	346.40	350	808.10
1,000	433.00	500	1,154.50

TABLE 2.1
HEAD IN METERS AND HEAD PRESSURE

Meters of Head	kPa	kPa	Meters of Head
1	10	5	.5
2	20	10	1
3	30	15	1.5
4	40	20	2
5	50	25	2.5
6	60	30	3
7	70	35	3.5
8	80	40	4
9	90	50	5
10	100	60	6
15	150	70	7
20	200	80	8
25	250	90	9
30	300	100	10
40	400	200	20
50	500	300	30
60	600	400	40
70	700	500	50
80	800	600	60
90	900	700	70
100	1000	800	80
200	2000	900	90
300	3000	1000	100

Normal Operating Pressure

The difference between static pressure and normal operating pressure is the friction caused by water for normal domestic and industrial consumption flowing through the various pipes, valves, and fittings in the system. As soon as water starts to flow through a distribution system, static pressure no longer exists. Flow varies with water consumption demands that change continuously during the day and night. Pressure readings on a test gauge attached to a hydrant with its main valve open, but not discharging water, indicate the normal operating pressure.

A hydraulic definition of normal operating pressure could read: "NORMAL OPERATING PRESSURE IS THAT PRESSURE FOUND ON A WATER DISTRIBUTION SYSTEM DURING NORMAL CONSUMPTION DEMANDS." Normal operating pressure is generally considered to be "STATIC PRESSURE" for fire flow test purposes since it represents the pressure available before water for fire fighting is flowing.

Residual Pressure

The word "residual" means a remainder or that which is left. During a fire flow test, the term residual pressure represents the remaining pressure on a distribution system within the vicinity of one or more flowing hydrants. The difference between normal operating pressure and residual pressure is the additional friction loss in the system caused by water flowing from the hydrants. Residual pressure will vary depending upon the water consumption demands. An important point is that residual pressure must be measured at the same location where the static reading is taken.

A hydraulic definition of residual pressure is as follows: "RESIDUAL PRESSURE IS THAT PART OF THE TOTAL THAT IS NOT USED TO OVERCOME FRICTION OR GRAVITY WHILE FORCING WATER THROUGH FIRE HOSE, PIPE, FITTINGS, AND ADAPTERS." The term residual pressure is also used to represent the pressure remaining in the water system while water is flowing for test purposes or fire fighting.

Flow Pressure

The rate of flow or velocity of the water coming from a discharge opening produces a pressure that is called flow pressure or velocity pressure. Since the stream of water from the discharge opening is not encased within a tube, it does not exert an outward pressure. The forward velocity of flow pressure can be measured by using a pitot tube and gauge. If a pitot tube is not available, a pressure gauge on a capped 2½-inch (65 mm) outlet opposite the flowing outlet may be used to measure flow pressure (Figure 2.12). This second method, though not technically as accurate, yields very close and usable approximations.

Figure 2.12 When a pitot tube is not available, flow pressure can be measured by placing a cap gauge on one 2½-inch (65 mm) outlet and discharging water from the other. *Courtesy of Edmond, Oklahoma Fire Department.*

A hydraulic definition of flow pressure is: "FLOW PRESSURE IS THE RECORDED FORWARD VELOCITY PRESSURE OF A FLUID AT A DISCHARGE OPENING."

The definitions of static, flow, residual, and normal operating pressure can be applied to the condition shown in Figure 2.13. This example represents a single 8-inch (200 mm) line, 1,000 feet (300 m) long that has an elevated storage tank at one end, two hydrants at the other end, and consumers between the storage and hydrants.

By studying Figure 2.13, an understanding of the types of pressures discussed can be gained. The pressure at point A is 40 pounds per square inch (276 kPa). This pressure is exerted by the height of the water in the tank. If there was no flow throughout the system, this pressure could be found at any given point. As previously stated, this is only a theoretical value since the water in the system is never in a true static condition.

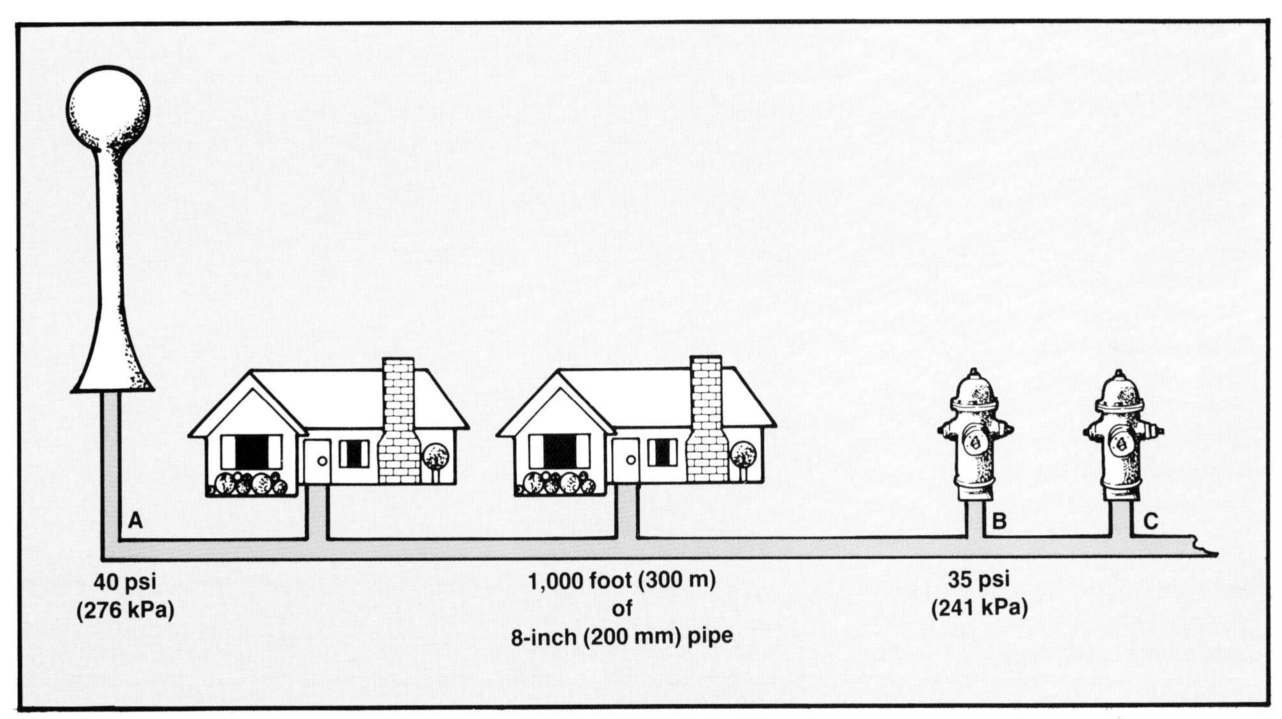

Figure 2.13 This system is static. The pressure measured at hydrant B is 35 psi (241 kPa).

If a pressure reading was taken at hydrant B or C with no other flow from the hydrant (a static reading), a pressure of somewhat less than 40 psi (276 kPa) would be obtained due to normal consumption. In the example, the pressure is found to be 35 psi (241 kPa) at hydrant B. This value is used as the static reading in the calculations, and it is the pressure that represents the potential energy available to force water to the hydrant.

If hydrant C was opened, a pressure drop on the gauge at hydrant B would occur. In the example, a pressure drop of 10 psi (69 kPa) has taken place, from 35 psi (241 kPa) to 25 psi (172 kPa)

(Figure 2.14). Thus, 25 psi (172 kPa) is the residual pressure, or pressure remaining in the main available to force water to other hydrants. If another outlet were opened at hydrant C or on another hydrant down the line, a pressure drop at hydrant B to a lower value would be noticed.

Flow pressure is obtained from an open outlet as at hydrant C. To obtain this pressure, a recording device such as a pitot tube or flowmeter must be used. In the example, a flow pressure of 10 psi (69 kPa) is observed. From this pressure reading, the gallon per minute (L/min) flow from the hydrant can be calculated. In this example, the flow would be 520 gpm (1 968 L/min).

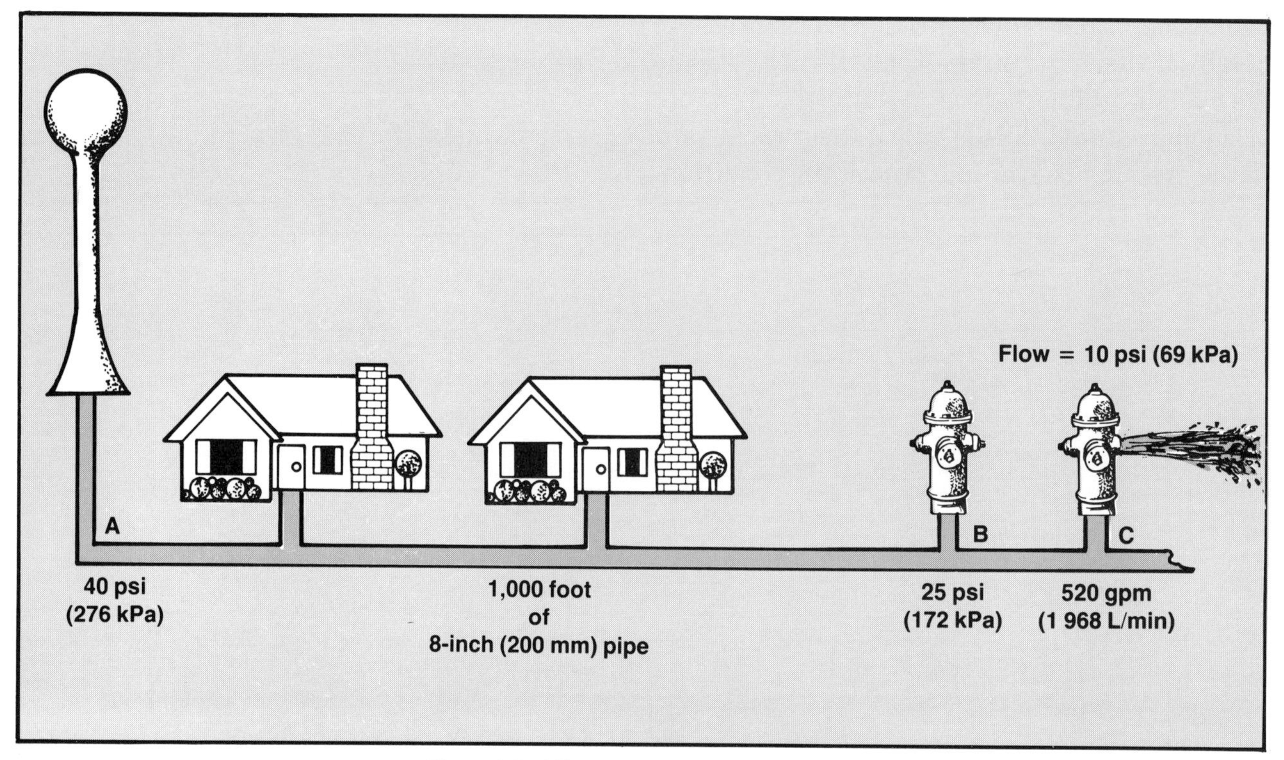

Figure 2.14 When hydrant C is opened, the pressure at hydrant B drops to 25 psi (172 kPa).

PRESSURE DEVELOPMENT

Pumps

Pressure on a water system may be created from the force imparted to water by pumps. There are several types of pumps utilized for water movement including reciprocating, rotary, centrifugal, and other miscellaneous types. The centrifugal pump is the most commonly used pump in water movement (Figure 2.15). The operation of a centrifugal pump is simply described as the drawing of liquid into the pump by gravity or creation of a vacuum. The liquid enters the eye of the impeller, it then turns 90 degrees as it enters vanes, and is thrown outward toward the inner walls of the casing. The water passes between the rim of the impeller and the casing and is ejected, under pressure, out the discharge port.

Figure 2.15 Centrifugal pumps are the most common type of pump in service today.

Elevated Storage

As discussed earlier, pressure is also created by the elevation of water above a water system. The developed pressure depends on the height of the tank. Elevated tanks may be used as a system pressure regulator or to handle peak surges on the system. The main purpose of elevated tanks is storage. A more detailed section on elevated storage can be found later in this chapter.

Pressure Tanks

Pressurized water tanks are generally used for private supply to sprinklers, standpipes, and hose systems. They may be seen occasionally supplying water for fire protection in rural schools, businesses, and similar occupancies. These tanks generally range from 1,000 to 10,000 gallons (3 785 L/min to 37 854 L/min) of storage. Tanks are filled two-thirds with water and the remaining third is charged to a 75 psi (517 kPa) minimum with air. Pressure tanks are not generally a direct concern of the fire department, but it should be realized that they will supply a limited quantity of water.

Flow Capacity of Water Mains

The ability of a water system to deliver an adequate quantity of water for fire fighting, domestic consumption, and industrial uses depends upon the carrying capacity of the system's network of pipes. When water flows through pipes, its movement causes friction which results in a pressure loss. The friction loss in pipe is similar to friction loss in fire hose.

Factors that affect the carrying capacity of water pipe include the size and length of pipe, the pressure at the source of supply, and the resistance to flow. Resistance is caused by elevation, internal friction, bends or turns in the pipe, joints, control valves, and other devices.

To better understand the flow carrying capacity of water mains, the piping C-factor must be defined. C is a factor indicating the coefficient of roughness of the interior of the pipe. The C-factor reflects the roughness of the inner surface of the piping and will decrease as the sediment, incrustation, and tuberculation within the pipe increase the obstruction to flow.

The effect that size and length has on the carrying capacity of pipe has been determined by experiments using an unlined cast iron pipe with no abnormal incrustations or deposits on the interior surface (C = 100). Using a static pressure of 50 psi (345 kPa) and a residual pressure of 20 psi (138 kPa), a 4-inch (100 mm) pipe 1,000 feet (300 m) long will only deliver approximately 255 gpm (965 L/min) while an 8-inch (200 mm) pipe 1,000 feet (300 m) long will flow around 1,575 gpm (5 962 L/min) (Figure 2.16). Many cities have adopted a policy prohibiting installation of 4-inch (100 mm) pipe because of the relatively small volume of water it will deliver.

The calculations for Figure 2.16 were derived using the Hazen-Williams formula. The Hazen-Williams formula was developed by experience and experiment based on the fact that the flow of water through circular pipe is governed by hydraulic law.

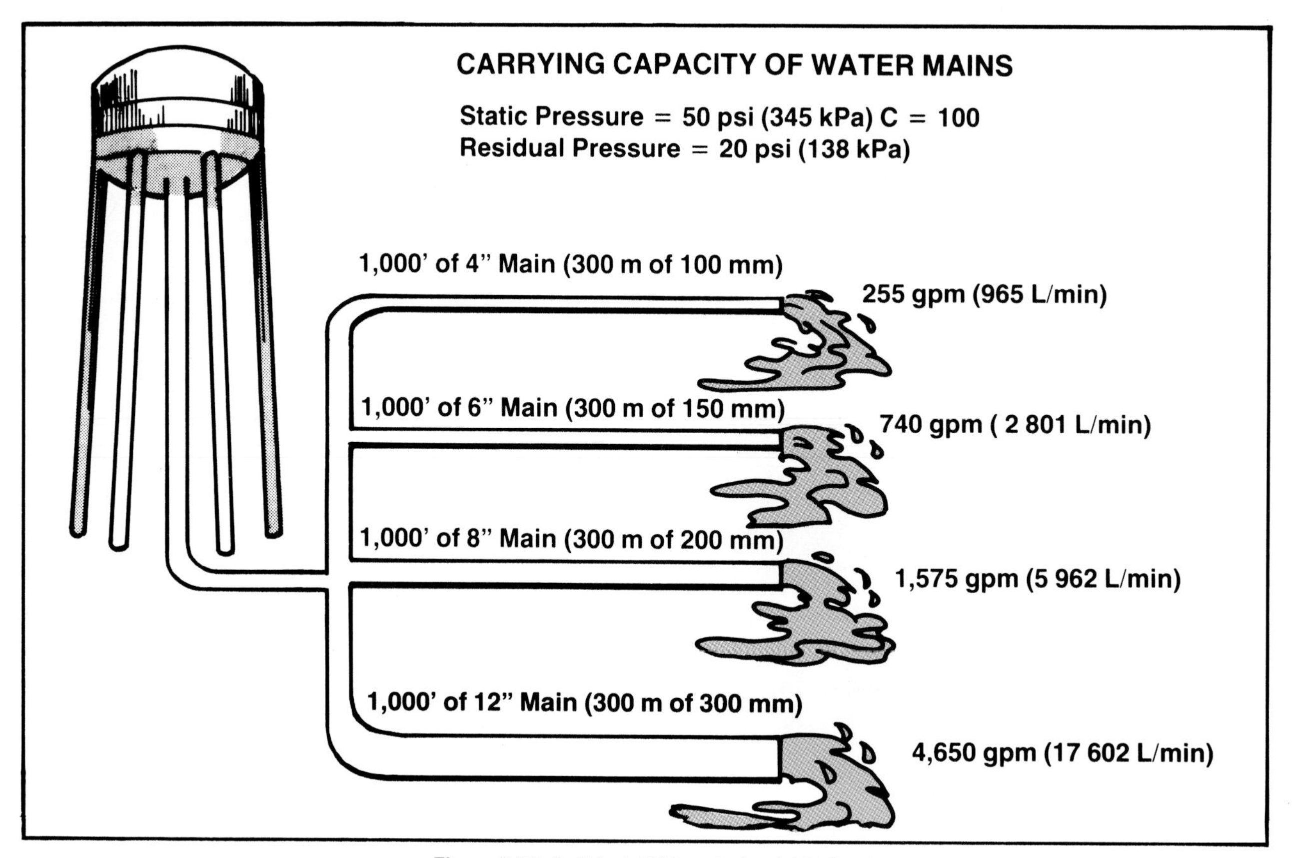

Figure 2.16 An 8-inch (200 mm) pipe 1,000 feet (300 m) long can deliver more than six times as much water as the same length of 4-inch (100 mm) pipe.

The quantity of water that is being used for domestic and industrial purposes materially affects the ability of a distribution system to maintain adequate pressure for fire protection. The rate of industrial water usage may be fairly constant throughout the year; however, the changing seasons may cause large differences in domestic water usage. On an extremely hot day, it is not unusual for maximum domestic use to be at least twice the daily average. For this reason, the ability of the distribution system to maintain adequate pressures must be considered at a time of maximum use of water for other than fire protection purposes.

For the convenience of comparison in the carrying capacities and friction loss in water pipe, Table 2.3 on page 46 shows the relative carrying capacities of pipes flowing full. This table is based upon the hydraulic law that the quantity of water carried by pipes of the same length and smoothness of surface, with a given loss of pressure, varies as the square roots of the fifth powers of the diameters. For example: "How much water will a 12-inch (300 mm) pipe carry as compared with a 6-inch (150 mm) pipe, with the same loss of pressure?" Locate 12 (300) in the left-hand column and 6 (150) in the vertical column. Where the two meet is 5.7. This shows that a 12-inch (300 mm) pipe will carry 5.7 times as much water as a 6-inch (150 mm) pipe. By the same method, it can be found that an 8-inch (200 mm) pipe will supply more than two 6-inch (150 mm) pipes.

For additional information, nomographs, and charts pertaining to the Hazen-Williams formula, refer to the National Fire Protection Association's *Fire Protection Handbook.*

Where a new hydrant is to be installed, the supply piping should be large enough to supply a minimum of 500 gpm (1 893 L/min) in addition to the domestic demands. Greater flow capacities, however, are recommended. If it is necessary to make a dead-end extension from a water system, care should be exercised to use a pipe of adequate size to supply sufficient water for fire protection in addition to domestic demands. Table 2.2 shows the maximum possible length of a dead-end for various sizes of

TABLE 2.2
MAXIMUM LENGTH OF DEAD-END SUPPLY TO HYDRANTS

Pipe Size	Capacity of Fire Department Pumper		
	500 GPM (1 893 L/min)	750 GPM (2 839 L/min)	1,000 GPM (3 785 L/min)
4-inch (100 mm)	190 feet (58 m)	90 feet (27 m)	50 feet (15 m)
6-inch (150 mm)	1,400 feet (427 m)	650 feet (198 m)	380 feet (116 m)
8-inch (200 mm)	5,600 feet (1 707 m)	2,650 feet (808 m)	1,550 feet (472 m)
10-inch (250 mm)	16,550 feet (5 044 m)	7,750 feet (2 362 m)	4,600 feet (1 402 m)
12-inch (300 mm)	40,500 feet (12 344 m)	19,300 feet (5 883 m)	11,150 feet (3 399 m)

pipe to supply various flows assuming a static pressure of 40 psi (276 kPa) and a residual pressure of 20 psi (138 kPa).

If fire hydrants are supplied by a circulating system, the flows will be greatly increased over the tabulation shown in Table 2.2. However, it should be pointed out that seldom can more than one pumper be used effectively where fire hydrants are supplied by 4-inch (100 mm) pipe, even if the main is circulating.

Static pressure has a definite effect on the carrying capacity of pipe. A residual pressure of 20 psi (138 kPa) is needed where the hydrant branch line connects to the water main. This residual pressure is necessary to overcome friction loss in the hydrant branch, in the hydrant, and in the intake hose of the fire department pumper. In Table 2.3 a static pressure of 40 psi (276 kPa) was assumed; by subtracting the needed 20 psi (138 kPa) residual pressure it leaves 20 psi (138 kPa) available to overcome friction in the pipe. If the static pressure were 50 psi (345 kPa) rather than 40 psi (276 kPa) there would be 30 psi (207 kPa) available for friction loss; thus, all possible lengths of pipe in the table would be 50 percent longer.

TABLE 2.3
APPROXIMATE DISCHARGE CAPACITIES OF PIPES FLOWING FULL

Diameter in Inches (mm)	Diameter in Inches (mm)									
	4 (100)	6 (150)	8 (200)	10 (250)	12 (300)	16 (400)	20 (500)	24 (600)	30 (750)	36 (900)
36 (900)	....	....	....	24.6	15.6	7.6	4.3	2.8	1.6	1
30 (750)	....	....	27.2	15.6	9.9	4.8	2.8	1.7	1	....
24 (600)	....	32.0	15.6	8.9	5.7	2.8	1.6	1	....	....
20 (500)	55.9	20.3	9.9	5.7	3.6	1.7	1	....	....	....
16 (400)	32.0	11.7	5.7	3.2	2.1	1	....	....	....	....
12 (300)	15.6	5.7	2.8	1.6	1	....	....	....	....	....
10 (250)	9.9	3.6	1.7	1	....	....	....	....	....	....
8 (200)	5.7	2.1	1	....	....	....	....	....	....	....
6 (150)	2.8	1	....	....	....	....	....	....	....	....

Table 2.3 shows the approximate relative carrying capacities of pipes flowing full. To compare a 12-inch (300 mm) pipe with a 6-inch (150 mm) pipe, locate 12-inches (300 mm) in the left column and go over until you find the multiplier under the 6-inch (150 mm) column (5.7). This means that a 12-inch (300 mm) pipe will flow 5.7 times as much as a 6-inch (150 mm) pipe.

Another factor that affects the carrying capacity of pipe is its resistance to the flow of water. The internal surface of pipe, regardless of the material from which it is made, offers resistance. New steel and unlined cast iron pipe have approximately equal resistance. The resistance in cement-lined cast iron pipe, plastic pipe, and asbestos cement pipe is somewhat less than that of a new steel or cast iron pipe.

Fittings, such as elbows, tees, pipe joints, valves, and meters also cause friction loss. For example, a 6-inch (150 mm) standard short-radius elbow will cause as much friction loss as 16 feet (5 m) of straight, new cast iron pipe. An 8-inch (200 mm) gate valve, fully opened, will cause about the same friction loss as 4½ feet (1.4 m) of straight 8-inch (200 mm) pipe. If, however, the valve is one-half closed, the loss will equal to the loss in 110 feet (34 m) of new, straight 8-inch (200 mm) pipe.

Water System Consumption

There are three rates of consumption that are considered when designing and evaluating water systems. They establish a base to which required fire flows can be added in designing a system or determining its adequacy. The rates are as follows:

- AVERAGE DAILY CONSUMPTION (ADC) — The average of the total amount of water used each day during a one-year period.
- MAXIMUM DAILY CONSUMPTION (MDC) — The maximum total amount of water used during any twenty-four hour interval in a three-year period. (Unusual situations which may have caused an excessive use of water, such as refilling a reservoir after cleaning, should not be considered in determining the maximum daily consumption.)
- PEAK HOURLY CONSUMPTION (PHC) — The maximum amount of water used in any given hour of a day.

The maximum daily consumption is normally about one-and-one-half times the average daily consumption. The peak hourly rate will vary from two to four times a normal hourly rate. The effect these varying consumption rates will have on the ability of the system to deliver required fire flows will vary with the system design. Both maximum daily consumption and peak hourly consumption should be considered to ensure that the water supplies and pressure do not reach dangerously low levels during these periods, and that adequate water will be available in the event of fire.

SOURCES

Water supplies for industrial and domestic consumption can be obtained from either surface or ground water. Although most water systems are supplied from only one source, there are instances where both sources are utilized. In general, surface waters are characterized by softness, suspended solids, some color, and considerable bacterial contamination. Conversely, ground waters are generally characterized by high concentrations of dissolved solids and gases, low color, relative hardness, and low bacteria. Surface water supplies are subject to relatively rapid depletion (except coastal waters) during a drought, yet the same supply source can make a quick recovery as a result of heavy rains and water runoff. There is, however, considerable lag time between a drought period and a noticeable water shortage in a well or spring fed ground water supply. Likewise, it may be several months after rainfall starts before restoration of ground water supply to a well or spring is complete. Surface waters include coastal water, rivers, streams, lakes, and ponds. Ground water supplies are usually wells and springs.

Large rivers are a source of domestic water supply for many cities. There are inherent treatment problems in utilizing this source of water, however, due to the wide variation in water quality. Water quality in a river is affected by upstream pollution and rainfall variance. As a result, treatment and purification procedures are subject to constant review and change.

If a lake is large enough and has adequate water runoff control (rain and snow melt), very little treatment may be necessary before the water is suitable for public consumption. The quality of water is relatively uniform in lakes, and treatment procedures are fairly consistent. Wide temperature variations in any supply may alter treatment requirements. Although most lakes have a wide temperature range, this variation occurs at a slow rate and will have less effect on the treatment. In general, lake contamination is highest near the shore and in shallow water.

Oceans and coastal waters are generally not used for domestic water purposes due to the cost of desalinization. They do, however, often serve coastal industries for some commercial purposes and provide a limitless fire fighting supply for coast area departments.

An impounded water supply will differ very little from a small lake, especially if its supply source depends upon flood waters or if the watershed is uncontrolled. Impounded water is much better when it can be permitted to stand for a considerable period of time before usage. Occasionally, an impounded area can be constructed or arranged so that a pumping station can supply the reservoir with water from a stream or other source. In this situation, there is considerable control over the quality of water pumped into the impounding area. Controlled impounded water supply and the period of storage will improve the quality of the water to a point where minimum treatment is necessary.

Water wells and springs produce water of the same general quality. The quality of water coming from wells or springs is generally good since earth formations aid in purifying the water. Very little treatment is necessary for ground water to be suitable for human consumption. Springs are water-bearing strata that intersect the surface of the earth. A well, even at another location, might penetrate the same strata and obtain the same water. It is possible for an underground strata to have a hydrostatic head adequate to force the water above the ground surface. Wells of this nature are known as flowing or artesian wells. Excessive use of artesian supplies and excessive pumping of wells will eventually cause a drop in the water table. This condition may cause an adverse decrease in water availability for a geographic area.

TYPES OF SUPPLY SYSTEMS

In order for water to be conveniently available at a tap, faucet, or fire hydrant, it must be moved to the distribution piping system from the city's water source. There are basically three types of systems that can be used by a municipality to deliver water from the source to the usage point. They are: gravity systems, direct pumping systems, and combination systems.

Gravity System

Gravity systems are used in communities where the water source is located at an elevation higher than the city (Figure 2.17). In this system, water is used from a natural or man-made reservoir. Intakes are then run to a treatment plant or directly into the distribution system. The pressure created to distribute the water is directly proportional to the difference in elevation between the source and the city. Since .433 psi (3 kPa) is developed for every foot (0.3 m) of elevation, a reservoir located 250 feet (76 m) above the city would provide a static pressure of approximately 110 psi (758 kPa). This is the same principle often used by industry when they erect gravity tanks to provide water for

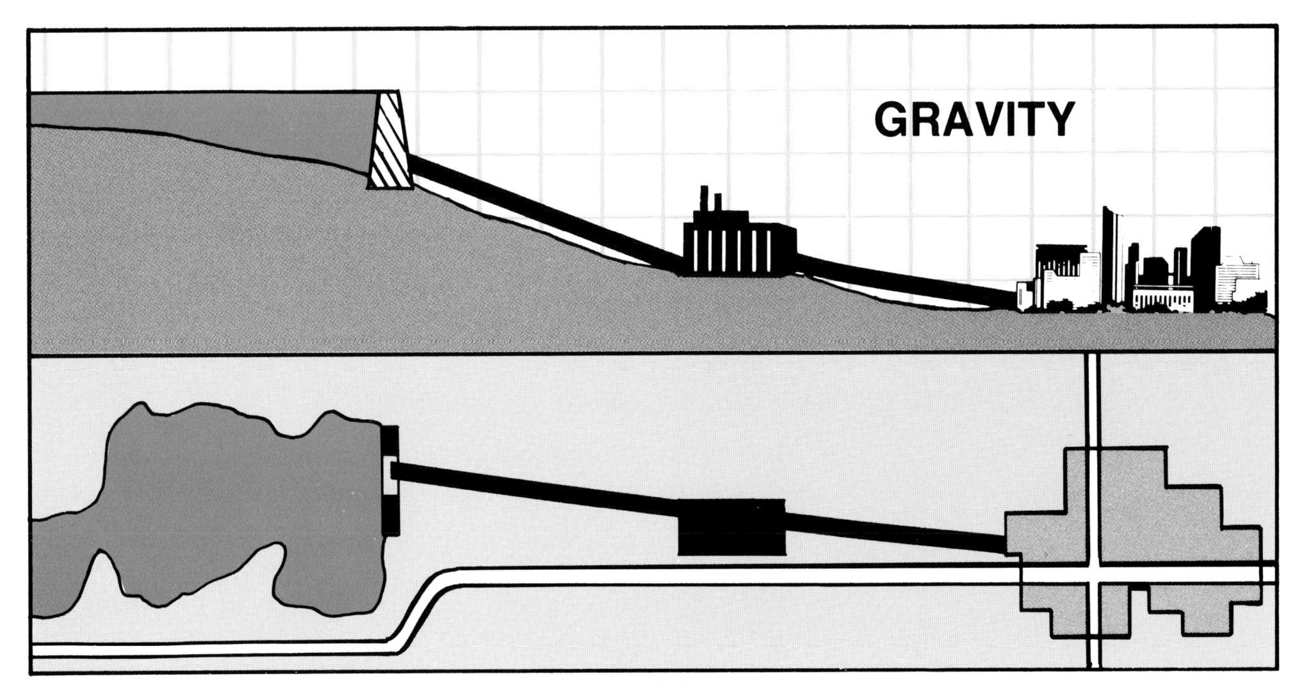

Figure 2.17 A gravity system is used where the water source is elevated.

sprinklers or a standpipe system. The important factors to consider when using a gravity supply are the following:

- Sufficient reservoir capacity and refill capability at maximum consumption
- Elevation difference great enough to assure adequate static and residual pressures in all parts of the distribution system

Direct Pumping System

In areas where the city source is located at or below the city's elevation, a direct pumping system may be used to supply water. Direct pumping systems may supply water to a treatment plant or pump water directly into the distribution system (Figure 2.18). These types of systems will have little or no elevated storage. The components in the direct pumping system include both the pumps and their prime movers or power supplies.

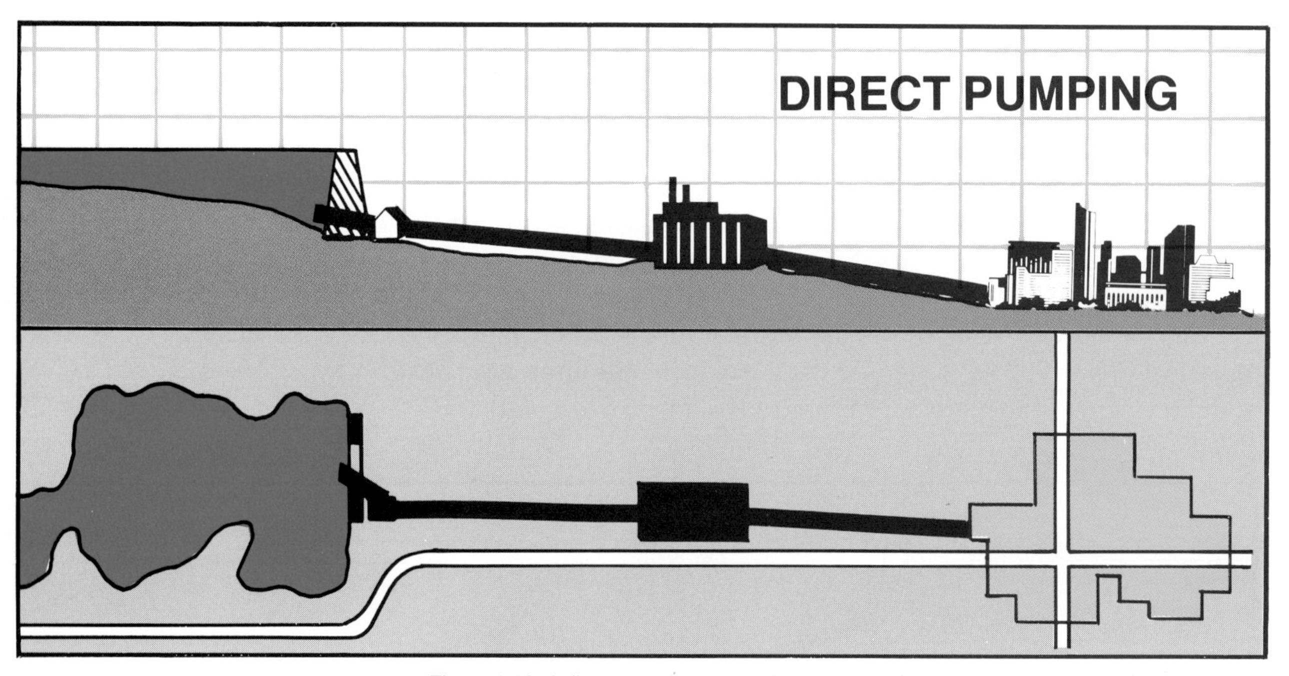

Figure 2.18 A direct pumping system is used when the water source does not have sufficient elevation to create adequate water pressure.

The type of pump and prime mover selected will depend on the type of use, the initial equipment cost, and the energy cost to keep it running over its expected life span.

Modern direct pumping systems may have computerized operations, regulating pumps, and mechanically automated valves. A major weakness with direct pumping systems, however, is the reliability of the system if a supply line, pump, or prime mover was to fail. Provisions to minimize the consequences of these failures include the following:

- Two or more adequate supply lines from the source

- Backup or auxiliary power supplies for all primary pumps
- Adequately supplied secondary pumping stations with a separate power supply and backup from the main electrical station

Combination System

This system is a combination of the gravity and direct pumping systems and is the type generally used in most communities. Water is pumped from the source to the treatment plant or directly into the distribution system. Standpipes, elevated tanks, or reservoirs are located at strategic points and serve as reserve storage containers for the system (Figure 2.19). These storage reservoirs are generally located in areas where the water system is most likely to need additional pressure and flow. These locations may include the ends of long feeder lines, extreme boundaries of the distribution system, or near areas that have exceptional peak demands.

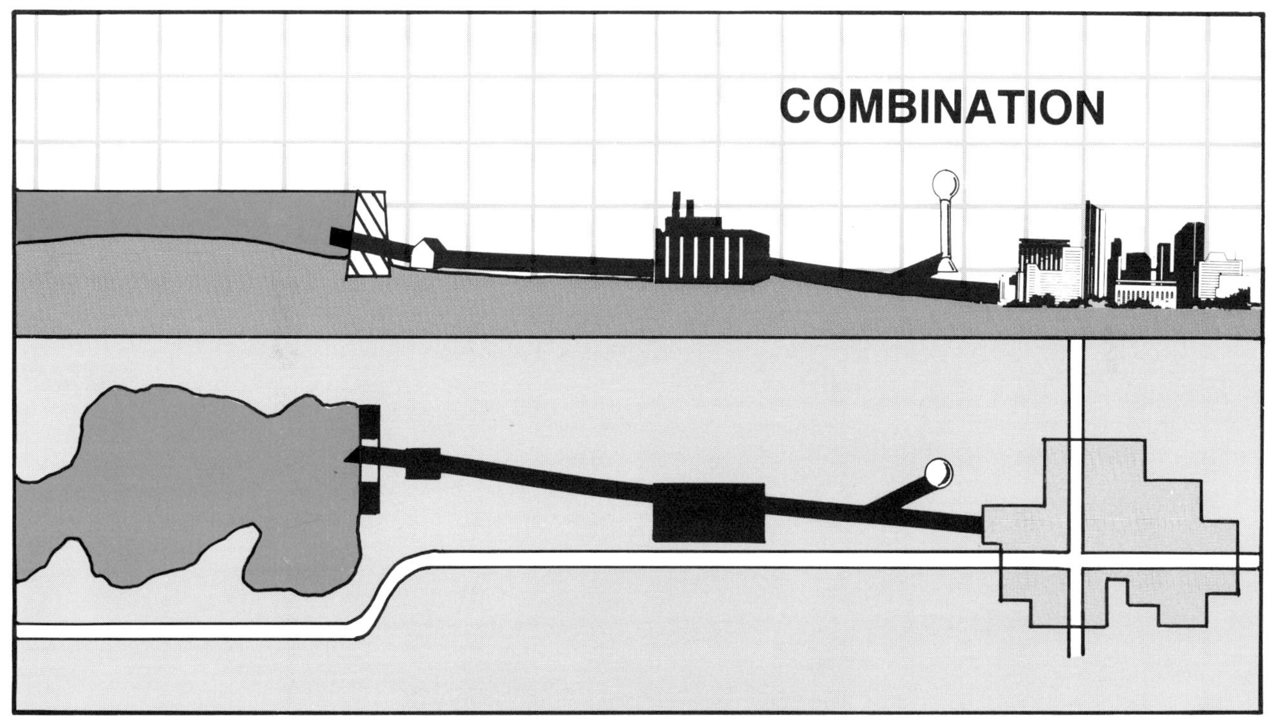

Figure 2.19 A combination of direct pumping and gravity is used to allow water storage during low demand; later, this water can be used when consumption exceeds pump capacity.

Within combination systems, water is pumped into the system at a reasonable uniform rate. When the demand is less than the rate at which the water is being pumped, the excess water automatically goes into an elevated storage tank or reservoir. When the consumption demand is greater than the amount supplied from the pumping station, water flows back into the system from storage. The size of supply pipe that is needed is determined by the maximum rate of flow required, pipe flow characteristics, and the available head pressure.

TREATMENT

Water supply treatment is a vital component in the delivery of a quality water supply (Figure 2.20). Treatment processes must deal with several factors including clarity, color, odor, contaminants, bacteria, and disease-carrying organisms. The Safe Drinking Water Act of 1974 provides federal and state requirements that impose specific limits of contamination which must be met. Treatment procedures for the purification of water include the following:

- Coagulation
- Sedimentation
- Filtration
- Chemical treatment of contaminants, bacteria, and organisms

Coagulation is used in conjunction with both filtration and sedimentation processes. Coagulants are chemicals that when added to the water cause small particles and contaminants to group together forming larger masses. In filtration, the increased size of the particles causes them to be trapped by the filter instead of allowing them to pass through. The increased weight of these masses aids sedimentation by starting or increasing the speed of the falling out process.

Figure 2.20 Water treatment facilities are generally located near the water source.

Sedimentation is the process of allowing materials to fall out of water by gravity force. For this to occur, water must be held relatively stationary in basins or filter beds. The settling material forms a sludge at the bottom of the basin and the basin is periodically cleaned to allow for continued settling.

Filtration is used to remove general waste and suspended matter by passing water through various types of filter media.

Filters used include slow, rapid, and pressure sand filters; varying sizes of metal screens; diatomaceous earth, and other materials.

Chemical treatment of water to destroy living organisms is generally done by chlorination, although other means are used. Other chemical treatments for water include

- Removal of calcium and magnesium (water softening)
- Fluoridation (decreased dental problems)
- Oxygenation or aeration

The concern of the fire department regarding a treatment facility is that a maintenance error, natural disaster, loss of power supply or fire could knock out pumping stations or severely hamper the purification process. This would drastically reduce the volume and pressure of water supplied to the municipality. Another factor that often weakens the supply of water to a community is if the treatment plant is unable to keep up with the demand during peak seasonal periods, primarily summer. This is the time of year when community water rationing is most common, and when the water volume may be drastically reduced in the distribution system.

DISTRIBUTION SYSTEMS

The elements within a water distribution system include pipes, valves, hydrants, meters, and other appliances for conveying water. These components are arranged in one of three layouts: a grid system, tree system, or circle or belt system. The most common layout is the grid system, but communities may be found with any one or a combination of the three types. Advantages of the grid system are that it is interlooped and connected at standard intervals, and through proper valving, can still supply the majority of the area if a main breaks (Figure 2.21 on next page).

Gridded systems contain the following types of supply piping:

- Primary Feeders — large pipes with relatively wide spacing that carry large quantities of water to various points of the system for distribution to smaller mains.
- Secondary Feeders — the network of intermediate size pipe that reinforces the grid by forming loops that interlock the primary feeders.
- Distributors — the smaller internal grid arrangements that serve consumer blocks and individual fire hydrants.

In the past, 4- and 6-inch (100 mm and 150 mm) pipe was extensively used for distribution in grids. Today, however, 8-inch (200 mm) pipe is fast becoming the minimum size installed. Many cities have limited the pipe size used in the distribution system to 8-, 12- or 16-inch (200 mm, 300 mm, or 400 mm) with no inter-

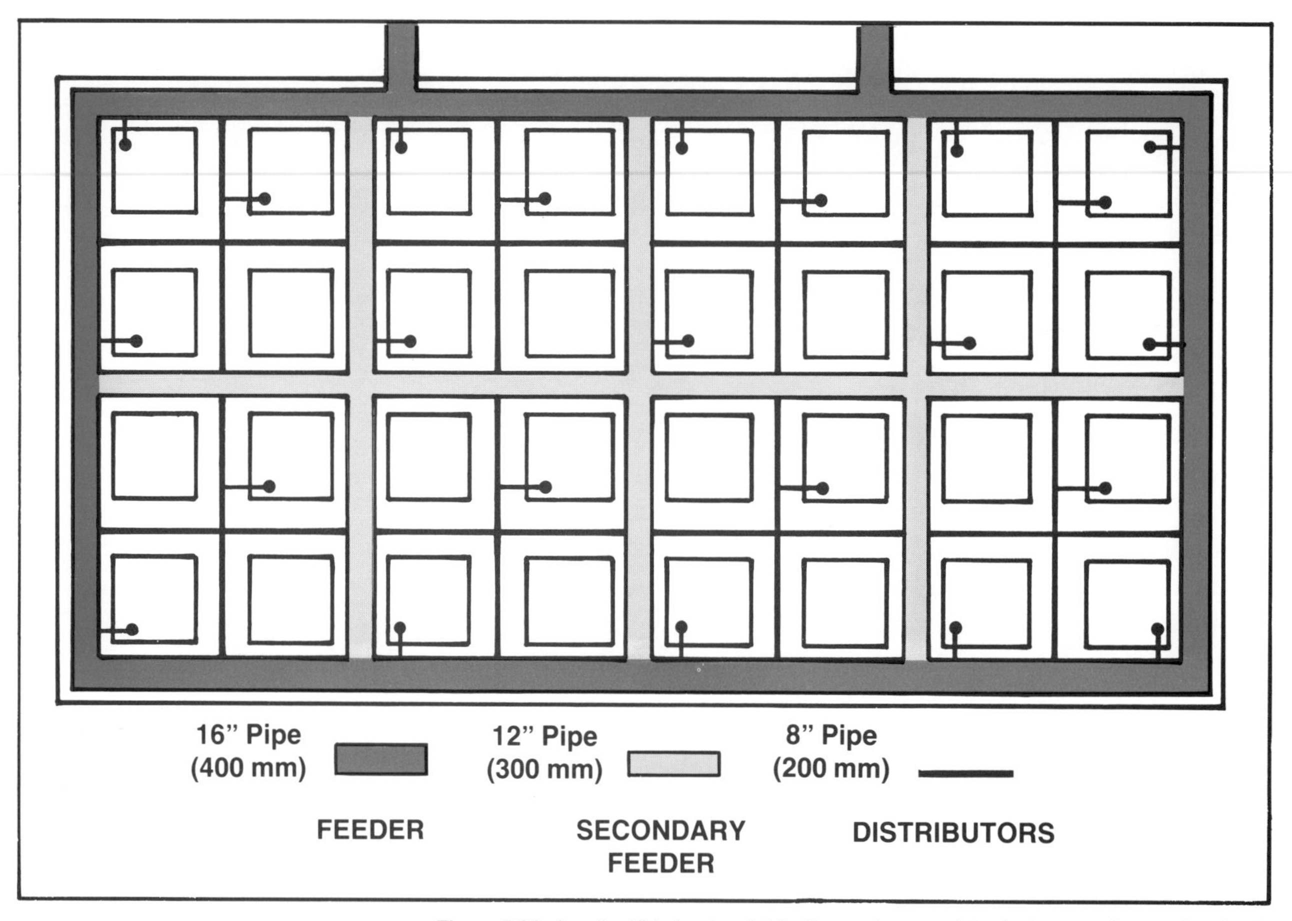

Figure 2.21 A well-gridded water distribution system consists of primary and secondary feeders and distribution mains.

mediate sizes permitted. Good fire protection practices suggest the following as the minimum size for pipe:

- Use 8-inch (200 mm) pipe for residential districts. Six-inch (150 mm) pipe to be used only where it will complete a good grid and is cross-connected at intervals not exceeding 600 feet (183 m).
- Use 8- and 12-inch (200 mm and 300 mm) pipe for shopping centers and industrial areas. Eight-inch (200 mm) pipe to be used only in sections where it will complete a good grid, and 12-inch (300 mm) for long lines not interconnected. Larger size mains may be needed depending upon size, layout, and occupancy of the structures.
- Use at least 8-inch (200 mm) pipe for multiple housing developments. Although, in many instances, parallel the requirement for industrial districts.

The distribution system may contain one or more pressure districts (zones or services). Multiple districts are usually found in larger cities or in those where considerable elevation differences are present. The purpose of separate pressure districts is to provide reasonable pressures for domestic, industrial, and fire demand throughout the distribution system. It may also prove necessary to increase pressures to properly supply hill areas or reduce pressure to areas of lower elevation (Figure 2.22). In larger cities with relatively flat terrain, it is often necessary to construct booster pumping stations to supply separate pressure districts. This arrangement provides for adequate pressure to areas located great distances from the source of supply.

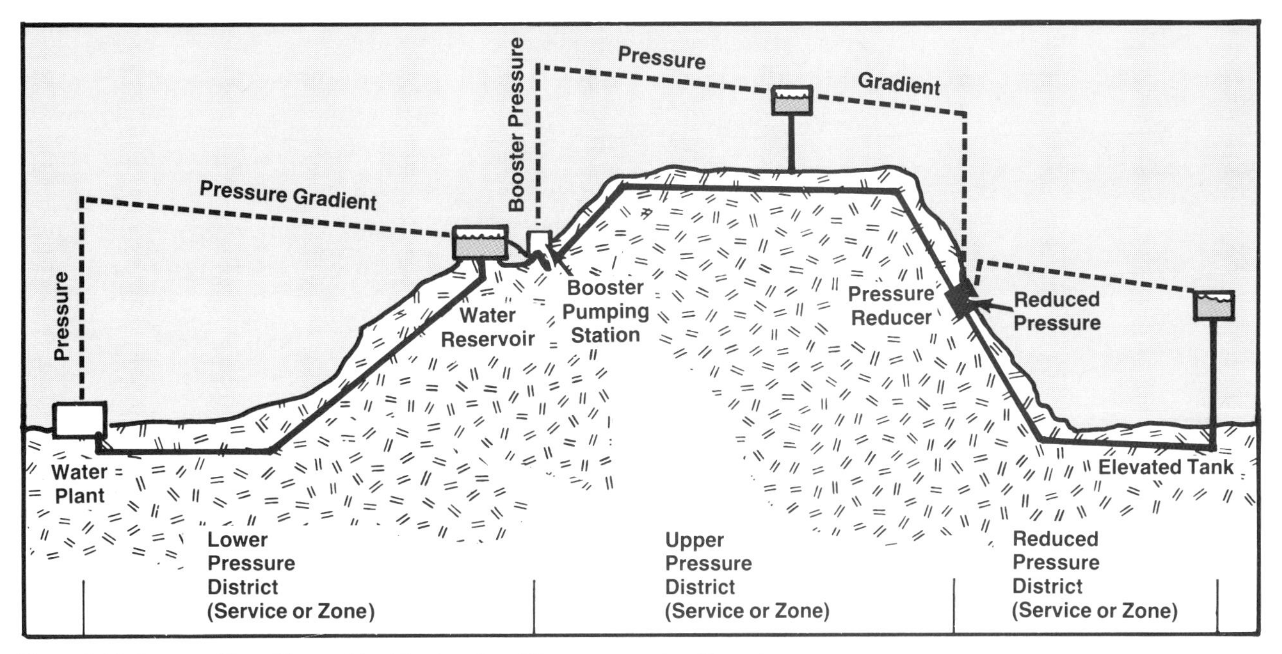

Figure 2.22 Communities with considerable elevation differences require different pressure districts or zones. This complicates water supply considerations.

Pumps

Virtually all water systems rely on pumps to serve as the water mover for distribution needs. The wide majority of pumps used for water movement are of the centrifugal-type including multistage, vertical turbine, propeller or axial flow, and submersible. The prime movers of these pumps may include diesel, gasoline, steam engines, or more frequently, electric motors. Pumps (and their prime movers) provide the motive force that sends water throughout the mains. They will be located in a system according to the needs of the consumers and the location of other system components such as tanks, treatment, and ground storage.

Storage

The storage of municipal water is performed for several reasons. The principal purposes include the following:

- A ready supply of potable water.
- Adequate capacity for normal and emergency use without service interruption.
- Maintain reasonably uniform pressure in the water system during high demand periods.

- Allow supply pumps sufficient cycle intervals, reducing wear and operating costs.
- Maintain storage water for fire fighting, thereby meeting insurance rating bureau requirements.
- Adequate service for weak areas within a water system.

Storage containers are classified in two types, ground storage reservoirs and elevated tanks. Ground storage reservoirs are generally found at treatment plants or wells and occasionally at high demand points on the distribution system (Figure 2.23). Ground storage tanks are built of steel or reinforced concrete (Figure 2.24). Pumping is generally used at these locations to maintain the desired pressure in the distribution system. However, ground storage reservoirs are also placed on hills and at high points in a system to utilize gravity flow. Elevated tanks are usually built of steel, but reinforced concrete is occasionally used (Figure 2.25). Elevated storage tanks are sometimes used at pumping stations to absorb pressure surges in the distribution system and at weak areas on the system to meet peak demands.

Figure 2.23 Open ground reservoirs are generally located at the water treatment plant and may be used as a water source for fire fighting operations.

Figure 2.24 Ground level storage tanks can supply large amounts of water for fire fighting operations. Due to their low elevation, they provide only a small amount of pressure, so it may be necessary to draft from the tank.

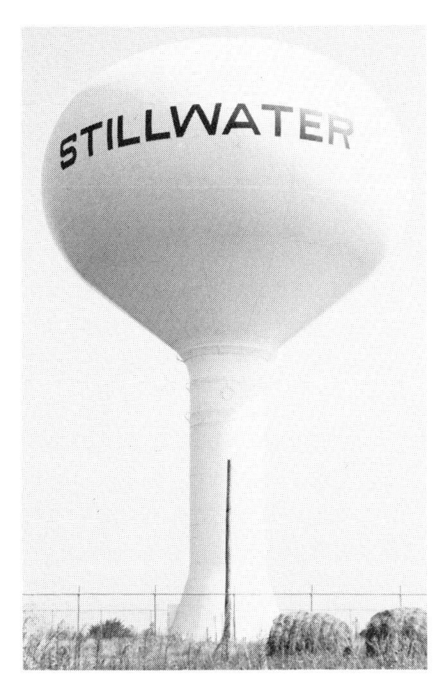

Figure 2.25 Elevated storage tanks are most commonly used as part of a combination water supply system. They can be filled during low demand periods and used to augment the pumping system during periods of high demand.

In some locations, elevated tanks are used solely for storage of fire protection water. For these tanks to be reliable, they must be properly located, have an adequate capacity, and be of sufficient height to develop the required pressures.

Elevated water tanks function in various methods depending on their use within a water system. Older tanks are generally open to the water system, and depending on the height of the water column, regulate the pressure in their area as well as controlling the cycle of pumps at respective pumping stations. Newer tanks have a hydraulically controlled mechanical valve that opens to fill the tank or supply water to the system, depending on the system pressure. These tanks may play a part in regular water system control or may be used solely for accommodating extremely high demand periods for domestic use or fire fighting.

The gravity flow from water tanks is generally reliable, but it should not be taken for granted. Tanks in extremely cold climates are subject to ice plugs that could block water flow and possibly damage the tank or riser. Water leaking from a damaged or poorly maintained tank or overflowing a tank in freezing weather can produce huge ice formations. This added weight can cause structural damage, riser pipes to be pulled loose, or even tank collapse. Tanks in cold climates may need to be outfitted with heating equipment to safeguard against these hazardous situations. The manufacturers or qualifed engineers should be consulted about the type and size requirements of a specific heating system.

A poorly maintained tank can lead to leaks and/or structural weakness that can severely hamper tank operation. Periodic cleaning and painting is required for steel tanks. All storage tanks should be inspected regularly to provide early detection of potential problems and required repairs.

Water Mains

A series of water mains forms the foundation of a good water supply. Size, location, tie-ins and loops, material, and proper maintenance all affect the quality and quantity of the water service delivered. Inadequate main size, deterioration due to age, corrroded lines, and poor maintenance can seriously hamper the fire department's ability since the water supply is not available. Fractures or breaks in major mains can cripple the water supply in a large area of a community. For this reason, communities must consider the quality, capability, and reliability of new system installations and retrofits. Fire department personnel can assist in these development areas by making recommendations concerning their needed requirements.

PIPE MATERIALS

Water mains are generally constructed of one of the following:

- Cast iron
- Ductile iron
- Steel
- Cement Asbestos
- Polyvinyl Chloride (PVC)
- Other plastics and synthetics

Cast iron was the earliest type of pipe material used and is still widely used today. It has good corrosion resistance and has a relatively long life. Some cast iron is lined with cement to improve its carrying capabilities.

Ductile iron maintains the properties of cast with the added mechanical properties of steel. It is generally used in areas where the mains are subject to shock and overhead loading.

Steel pipe is used for applications requiring large diameter pipe or in special situations. Steel pipe has good strength/weight characteristics but has low corrosion resistance.

Cement asbestos is another common type of material used for water mains. It has excellent corrosion resistance and good carrying characteristics, but is subject to damage from shock or external loading.

PVC and other plastic pipe is seeing increased use as water supply piping. It is quite possible that plastic pipe will become the standard for water distribution in the future. Plastic pipe is lightweight, has excellent corrosion resistance, and very good flow characteristics. One area as yet unknown is the expected lifespan of plastics since they have only recently been utilized.

PIPE JOINTS AND CONNECTION APPLIANCES

Pipe sections must be kept at lengths that allow for easy handling and installation. Pipe joints are used to connect these sections. The type of joints used will depend on the type of pipe and its intended use.

Joints for cast iron pipe may include bell-spigot, flanged, compression, mechanical, and threaded. Joints for other piping may include these types as well as special service joints, such as expansion or flexible types. Some common types of joints are shown in Figure 2.26.

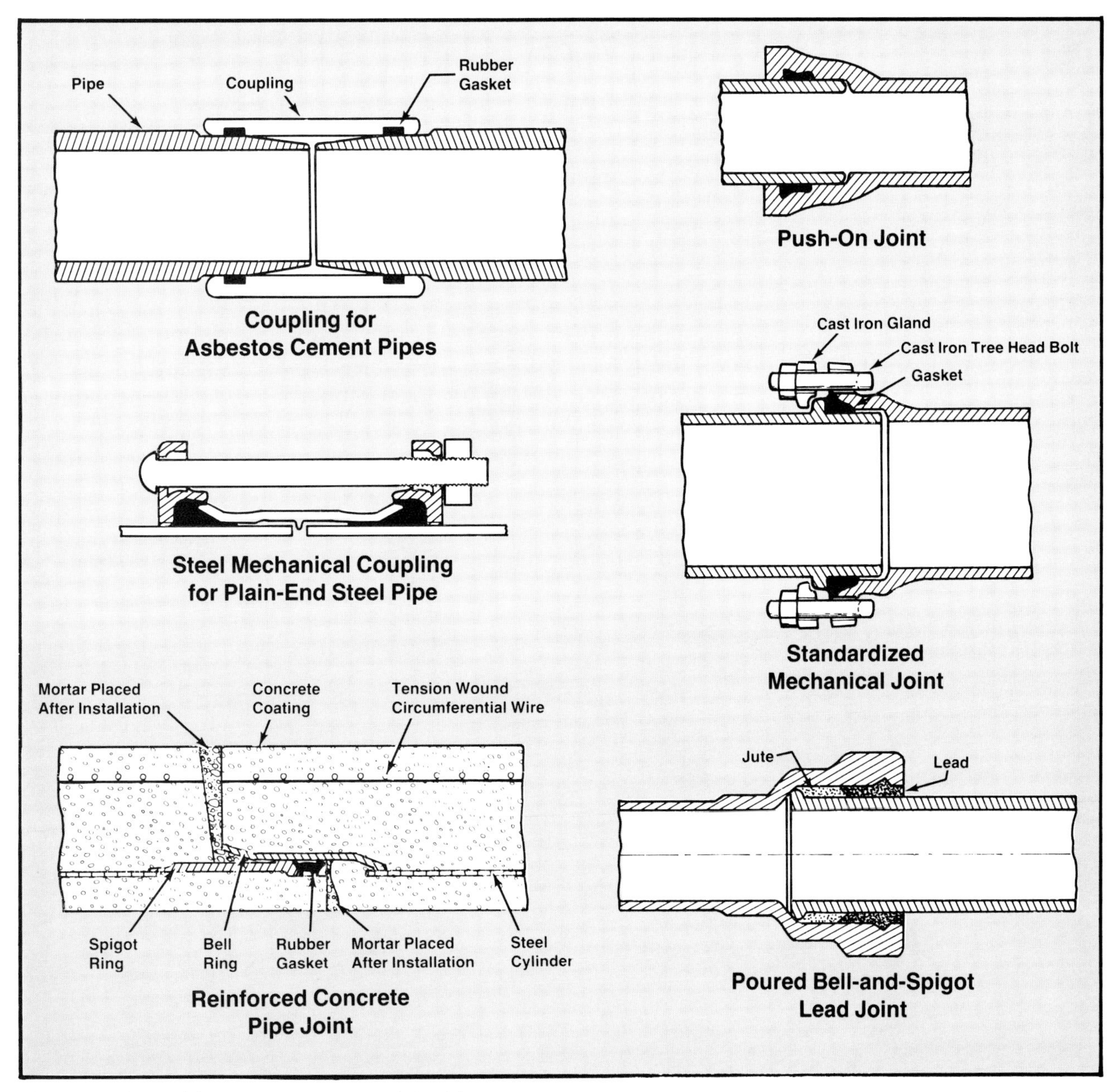

Figure 2.26 Some of the more common types of pipe joints.

Figure 2.27 A thrust block has been set directly beneath the hydrant riser.

Tees, elbows, and crosses are utilized to split or divert flow and to tie together sections of piping grids. Due to the force of water changing direction at these locations, concrete pourings called thrust blocks are set in place to prevent movement of the mains at these junctions (Figure 2.27).

The fire department should be concerned about the various types of connections used because they are often points of leakage and cause additional friction loss. For example, a 6-inch (150 mm) standard short radius elbow causes as much friction loss as 16 feet (5 m) of straight 6-inch (150 mm) new cast iron pipe. For this reason, the fire department should encourage the use of long radius elbows and low friction loss connectors.

Water Main Valves

The function of a valve in a water distribution system is to provide a means of controlling the flow of water through the distribution piping. Valves should be operated at least once a year to keep them in good condition. Valve spacing should be such that only a short length of pipe will be out of service at one time should a break occur. The maximum lengths between valves should be 500 feet (150 m) in high-value districts and 800 feet (244 m) in other areas.

Valves for water systems are broadly divided into indicating and nonindicating types. An indicating valve visually shows the position of the gate or valve seat (closed, partially closed, or open). Valves in private fire protection systems should be of the indicating type (Figures 2.28a and b). Except for possibly a few valves in treatment plants and pump stations, valves in public water systems are of the nonindicating type. Valves in a water distribution

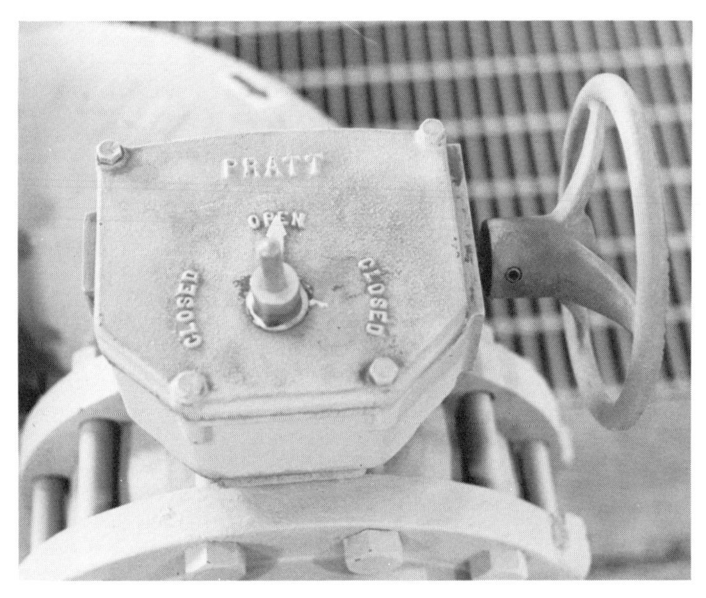

Figure 2.28a The markings on this valve clearly tell whether the valve is open or closed.

Figure 2.28b OS&Y valves are commonly used on water supply systems. When the stem is out, the valve is open. When the stem is not visible, the valve is closed.

system are normally buried or installed in manholes (Figure 2.29). If a buried valve is properly installed, a way to operate the valve from above ground through a valve box will be provided. A special socket on the end of a reach rod is known as a valve key or gate key and is used for this purpose (Figure 2.30). Control valves in water distribution systems may be either gate valves or butterfly valves. Gate valves are usually of the non-rising stem type and as the valve nut is turned by the valve key (wrench), the gate either raises or lowers to control the water flow. Butterfly valves usually have a rubber or rubber composition seat which is bonded to the valve body. The valve disk rotates 90 degrees from a fully open to tightly shut position. The valve's principle of operation provides satisfactory water control after long periods of inactivity. Both valves can be of the indicating and nonindicating type.

Figure 2.29 A nonindicating valve in position on a water main.

The advantages of proper valving in a distribution system are readily apparent. If valves are installed according to NFPA and/or AWWA established standards, only one or perhaps two fire hydrants will be closed while a single break is being repaired. The advantages of proper valving will, however, be reduced if all valves are not properly maintained and kept fully open. Very high friction loss is caused by partially opened valves. When valves are closed or partially closed, the condition may not be noticeable during ordinary domestic flows of water. Therefore, the impairment will not be known until a fire occurs, detailed inspections, and/or fire flow tests are made. A fire department will experience difficulty in obtaining water in areas where there are closed or partially closed valves in the distribution system.

The importance of a valve maintenance program cannot be overemphasized. Valves should be inspected after each use.

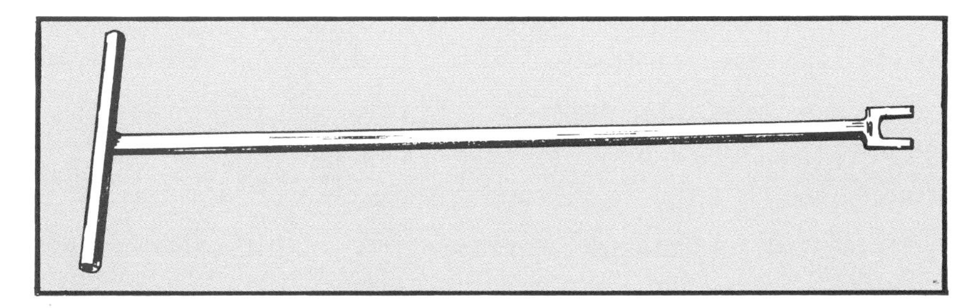

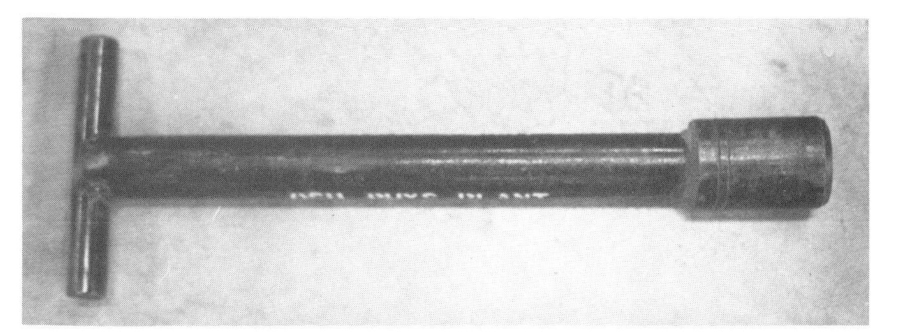

Figure 2.30 Valve wrenches for water main valves differ according to the type of valve. Without the proper wrench, it may not be possible to operate a valve.

Periodic inspections should also be made, and the valves operated annually. Annual fire flow tests will provide reliable assurance that all valves are open if the water flow is normal. In a grid distribution system, it is usually necessary to close more than one valve to stop the flow of water from a break and, in some cases, as many as four to six valves must be closed. As the necessary valves are closed, the flow is retarded but will not stop until the last valve is closed. After repairs are completed, a flow test should be made to verify that previous flow capacity is obtained and that all valves are opened. By failing to open a valve, it is possible that distribution piping will have a diminished flow rather than the full circulation flow. This valve positioning can be confirmed by conducting a flow test.

Proper maintenance of valves is much easier when all valves in a system have uniform direction of stem operation. Proper maintenance may sometimes be curtailed if valve boxes and manholes are not kept clean, their covers not maintained to grade, or if the valves have been covered with pavement. Rigid control must be maintained by the water department over the opening and closing of valves in the system. Proper records of valves and their locations are essential.

Fire Flowmeters/Regulators

Fire flowmeters are used on public water systems to measure water flow amounts into a given occupancy for normal use without obstructing flows for fire protection. These meters are generally found on connections between private fire protection systems and the public water system. Fire flowmeters are classified in two types: detector check valve meters and full registration meters.

Detector check valves bypass a small quantity of water through a disk meter that measures the smaller flows through this bypass meter. When heavy flows are demanded, a check valve opens and water flows straight through unmetered. Disk meters are available to three inches (77 mm).

Full registration meters are available in three basic types: displacement, proportional, or turbine. These units are designed for large flows with minimum friction loss.

The displacement-type meter has a bypass and disk meter similar to that in the detector check valve for small flows. When the valve trips, the first bypass closes and a second bypass is opened that flows a small quantity of water through a current meter that proportionally measures the open waterway flow. Total flow is the sum of both meters.

A proportional fire flowmeter has two measuring devices. Small flows are measured by a disk meter in the bypass. Large flows are measured by a proportional meter calibrated to record the total flow in the line, when the automatic valve opens the main line. The sum of both measuring devices gives the total flow through the meter.

Turbine-type meters consist of two parts: a main case and a measuring chamber. In the measuring chamber a rotor diverts water to a measuring chamber assembly where the register is calibrated to record total flow. A fire service strainer is recommended upstream from the meter.

Fire flowmeters measure high volume flow with a very low friction loss and should not be confused with domestic meters that measure small volumes and have a relatively high friction loss.

Local authorities and/or qualified engineers should be consulted about the installation of fire flowmeters since they may require special valving to prevent the private system water from entering the public system.

Pressure Regulating Valves

Water flowing from a high-pressure source (a reservoir at a high elevation, a high-pressure main, or a high-pressure zone) will require a downstream pressure regulatory device to prevent damage to water mains and piping. A pressure regulating valve is used to reduce the higher pressures.

These devices open and close automatically depending on flow demands. There are two types of valves: differential and pilot-operated fixed pressure. The differential valve sustains a constant pressure drop between the inlet and outlet for all flow rates. The pilot-operated fixed pressure valve maintains a constant outlet pressure at any flow rate within the valve's capacity.

The pressure reduction is caused by the friction loss in the valve opening. The friction loss varies with the position of the valve piston. As the pressure drops the valve opens further. The maximum friction loss is reported as similar to a globe valve of the same size.

For further information regarding the various appliances and connectors, refer to the appropriate AWWA specifications available from: American Water Works Association, 6666 West Quincy Ave., Denver, Colorado 80235.

FLOW OBSTRUCTIONS

Water mains are subject to several types of flow obstructions that can gradually or immediately limit the flow capacity of the main. Obstructions that are commonly found in water mains may be the result of one or more of the following:

- Incrustation
  - — tubercular corrosion or rust
  - — chemical constituents in the water
  - — growth of biological organisms

- Deposits (Figure 2.31)
  - — Sedimentary (mud, clay, pebbles, leaves, and organic decay)
- Foreign matter
  - — waste
  - — tools
- Malicious damage
- Valves

The three causes of incrustation as well as sedimentary deposits cause a gradual decrease in the mains carrying capacity and will be seen through evaluation of the fire flow tests at a given location over time. (Decreasing flow and higher pressure drops.) Foreign matter or problems with valves, however, may cause a dramatic drop in capacity and will be evident between test comparisons from year to year.

Incrustation is caused by tuberculation of the pipe that forms nodules on the interior lining and/or by the natural corrosion of unlined piping. It may also occur due to deposits left when certain organisms chemically react with the pipe metal. Sedimentary deposits caused by the fallout of mud, clay, or dead organisms will line the bottom interior of the pipe and will show the same gradual decrease in capacity. These factors are often taken into account by water supply engineers in the form of a decreasing C-factor. New cast iron pipe is assigned a C-factor of 120 while 15- to 20-year-old pipe is given a coefficient of 100.

Foreign matter, other than sediment, may include surplus lead which seeps or accidentally runs into a main when a bell and spigot joint in cast iron is poured with molten lead. Chunks of lead as large as 40 pounds (18 kg) have been reported found in water

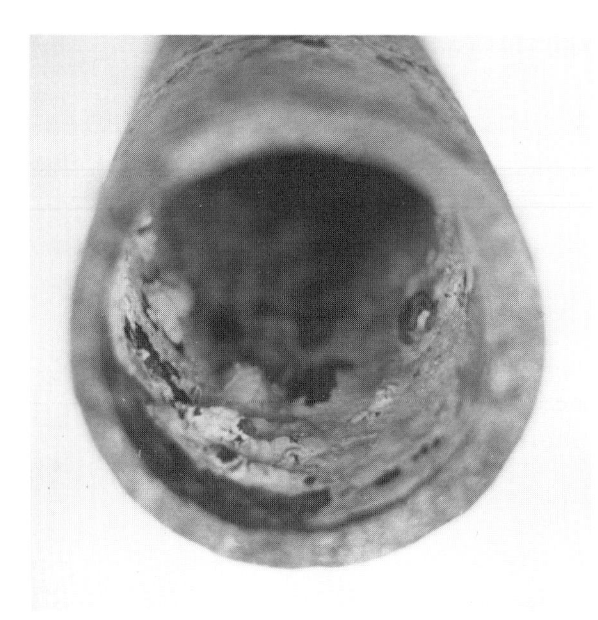

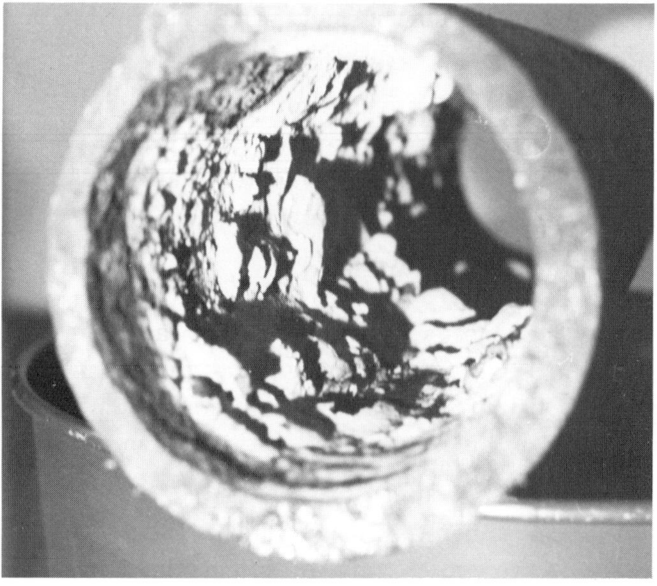

Figure 2.31 Examples of severe deposit buildup in water pipes.

mains. In other instances, boards, crowbars, pickhandles, stones, and various other obstructions have been found. It can be readily seen that the cause of increased friction loss and a decreased water flow might be a combination of two or three types of obstructions.

Fire department officials should further realize that the malicious obstruction of mains and connections are possible. Cans, bottles, paper waste, and other materials can be placed in mains to be installed or stuffed in open hydrant outlets or fire department sprinkler and standpipe connections. Proper tightening or replacement of connection and outlet caps will negate the majority of these problems.

Methods of removing the various kinds of obstructions have been developed and many water mains are successfully cleaned each year. One method of cleaning obstructions is by the flexible pipe cleaning tool as illustrated in Figure 2.32. Municipal water departments do not usually possess the equipment necessary to clean water mains. This work is usually contracted to companies who specialize in this type of work. Some waterworks personnel hesitate to clean pipelines mechanically because they fear problems with sediment and residues which may appear after cleaning. Methods are available to prevent the problem of organic growth if the chemistry of the water is determined. Little can be done, however, to prevent retuberculation of pipes unless the interior can be cement lined.

Chemical control of organic growth in pipelines can be performed if careful analysis of the water is made and the proper additives are used. Improper chemical control may cause a need for chemical cleaning.

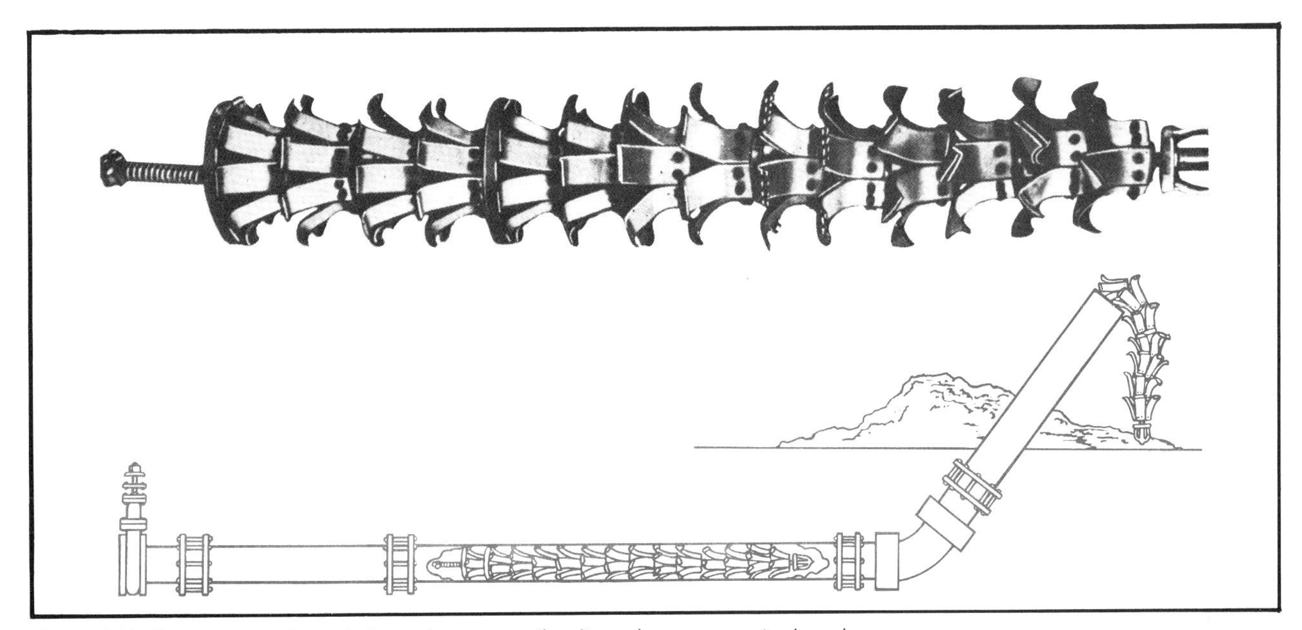

Figure 2.32 To restore a water main to maximum operation, it may be necessary to clean deposits from the pipe. Special pipe cleaning devices are used for this operation.

Chapter 2 Review

Answers on page 260

TRUE-FALSE: Mark each statement true or false. If false, explain why.

1. Flow pressure is the recorded forward velocity pressure of a fluid at a discharge opening.

   ☐ T ☐ F ______________________________

2. The positive displacement pump is the most common type of pump used in water movement today.

   ☐ T ☐ F ______________________________

3. Industrial water usage fluctuates more with the changing seasons than does domestic water usage.

   ☐ T ☐ F ______________________________

4. The three basic types of pipe system layouts are the grid system, the tree system, and the combination system.

   ☐ T ☐ F ______________________________

5. The two basic types of water storage containers are ground containers and elevated containers.

   ☐ T ☐ F ______________________________

6. During a water main break, the flow of water can usually be controlled by closing just one valve.

   ☐ T ☐ F ______________________________

7. The pressure exerted on the bottom of a container by a liquid is dependent upon the shape of the container.

   ☐ T ☐ F ______________________________

MULTIPLE CHOICE: Circle the correct answer.

8. Which one of the following is *not* a pressure found with liquids?
   A. Atmospheric
   B. Static
   C. Hydrant
   D. Flow

9. What is the minimum desired flow for a new hydrant installation?
A. 250 gpm or 946 L/min
B. 500 gpm or 1 893 L/min
C. 750 gpm or 2 840 L/min
D. 1,000 gpm or 3 785 L/min

10. The two basic sources of water for supply systems are ground water and ________.
A. sea
B. river
C. hydrant
D. surface

11. Which one of the following is *not* a major type of water flow obstruction?
A. Incrustation
B. Tubercular corrosion or rust
C. Chemical constituents in the water
D. Small animals

SELECT: Circle the correct response.

12. Gravity systems are used when the water supply is higher/lower than the city.

SHORT ANSWER: Answer each item briefly.

13. Pressure is expressed in what unit of measure?

____

____

14. What is the weight of 1 cubic foot (0.1 m^3) of water?

____

____

15. What is the pressure exerted by a 1-foot (0.3 m) column of water?

____

____

16. Briefly describe the operation of a centrifugal pump.

____

____

17. Name the three basic rates of water consumption.
A. ____________________
B. ____________________
C. ____________________

18. Name the three basic types of supply systems used by a municipality to deliver water.

A. ______________________________

B. ______________________________

C. ______________________________

19. Name at least three elements within a water distribution system.

____

____

____

20. Name three principal purposes for the storage of municipal water.

____

____

____

21. Name three conditions of water mains that may hamper the fire department's ability to utilize the water from that main.

____

____

____

22. Name at least three materials from which pipes are constructed.

____

____

____

23. How often should valves be inspected?

____

____

____

Photo by Joel Woods

Chapter 3

Fire Hydrants

This chapter provides information that addresses performance objectives in NFPA 1001, *Fire Fighter Professional Qualifications* (1987) and NFPA 1002, *Fire Apparatus Driver/Operator Professional Qualifications* (1988), particularly those referenced in the following sections:

NFPA 1001

Fire Fighter II

4-15.3

4-15.6

NFPA 1002

3-2.2

3-2.8

Chapter 3
Fire Hydrants

Early water systems used hollowed out logs for water mains. The method of obtaining water for fire fighting purposes was crude: pits were dug at specified intervals to expose the mains. A hole was then made in the main and a wooden plug was inserted (Figure 3.1). These plugs were known as fire plugs and this term is still commonly used to identify fire hydrants. When a fire occurred, the plug was removed allowing water to fill the pit. Fire apparatus took draft from the pit; however, the flow of water was usually so meager that the system was seldom an effective aid to fire fighting.

Eventually, the use of cast iron pipe permitted system pressures to be increased, and this practice led to the development of the post-type fire hydrant. An opening or hose connection at the upper end of the standpipe provided a place from which fire pumps received their supply. Compared to the "post-type" hydrant, it

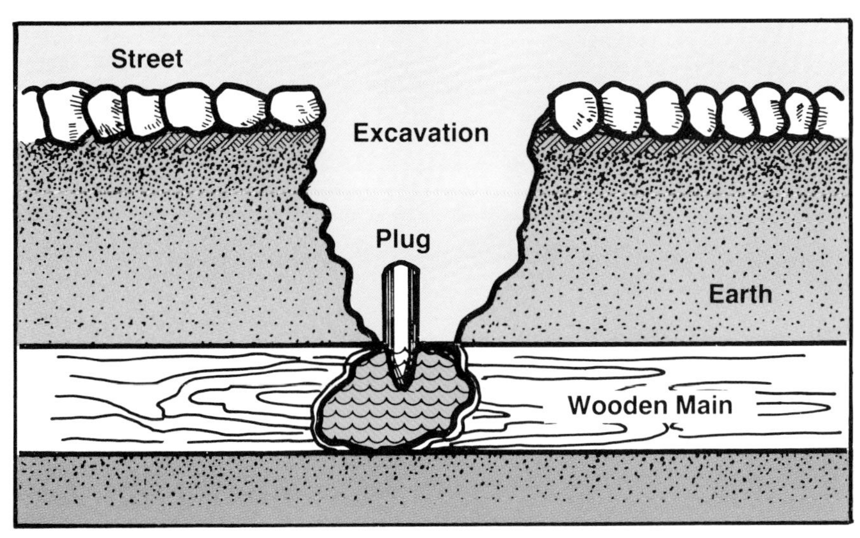

Figure 3.1 The first fire hydrants consisted of plugs that were driven into wooden water mains.

Figure 3.2 Early post-type fire hydrants were made of wood and held together by metal strips.

was soon realized that the "plug-type" hydrant had definite limitations and was practically useless under higher pressures.

The first post-type fire hydrant in the United States was designed around 1800. It had both a hose connection and a faucet. It also had a barrel that was always charged with water because the valve was in the top. Freezing weather caused considerable trouble and, for this reason, the hydrant was later covered by an octagon-shaped wooden enclosure held together by iron strips (Figure 3.2). An improved hydrant, with the valve at the base and the facilities to drain the barrel, was later introduced. Hydrant design continued to improve and the hydrants that followed were, for the purpose of general description, similar to our present-day fire hydrants (Figure 3.3).

Figure 3.3 Several types of modern fire hydrants.

TYPES OF FIRE HYDRANTS

The two main types of modern hydrants are DRY-BARREL and WET-BARREL. The dry-barrel operates with a compression valve opening against the pressure or a knuckle-joint type opening with the pressure. Generally, any water that remains in a closed dry-barrel hydrant will drain through a small orifice at the bottom. This drain opens as the main valve approaches the closed position. These hydrants are used in areas that are subject to freezing weather conditions and are by far the most common hydrants in use today. However, in some areas the drains are being plugged to prevent contamination from the water table. Hydrants then must be pumped out after use.

Wet-barrel hydrants usually have a compression-type valve at each outlet, or they may have only one valve located in the bonnet that controls the flow of water to all outlets. Typical dry-barrel and wet-barrel hydrants are shown in Figure 3.4. In general, hydrant bonnets, barrels, and foot pieces are made of cast iron. The important working parts are usually made of bronze, but

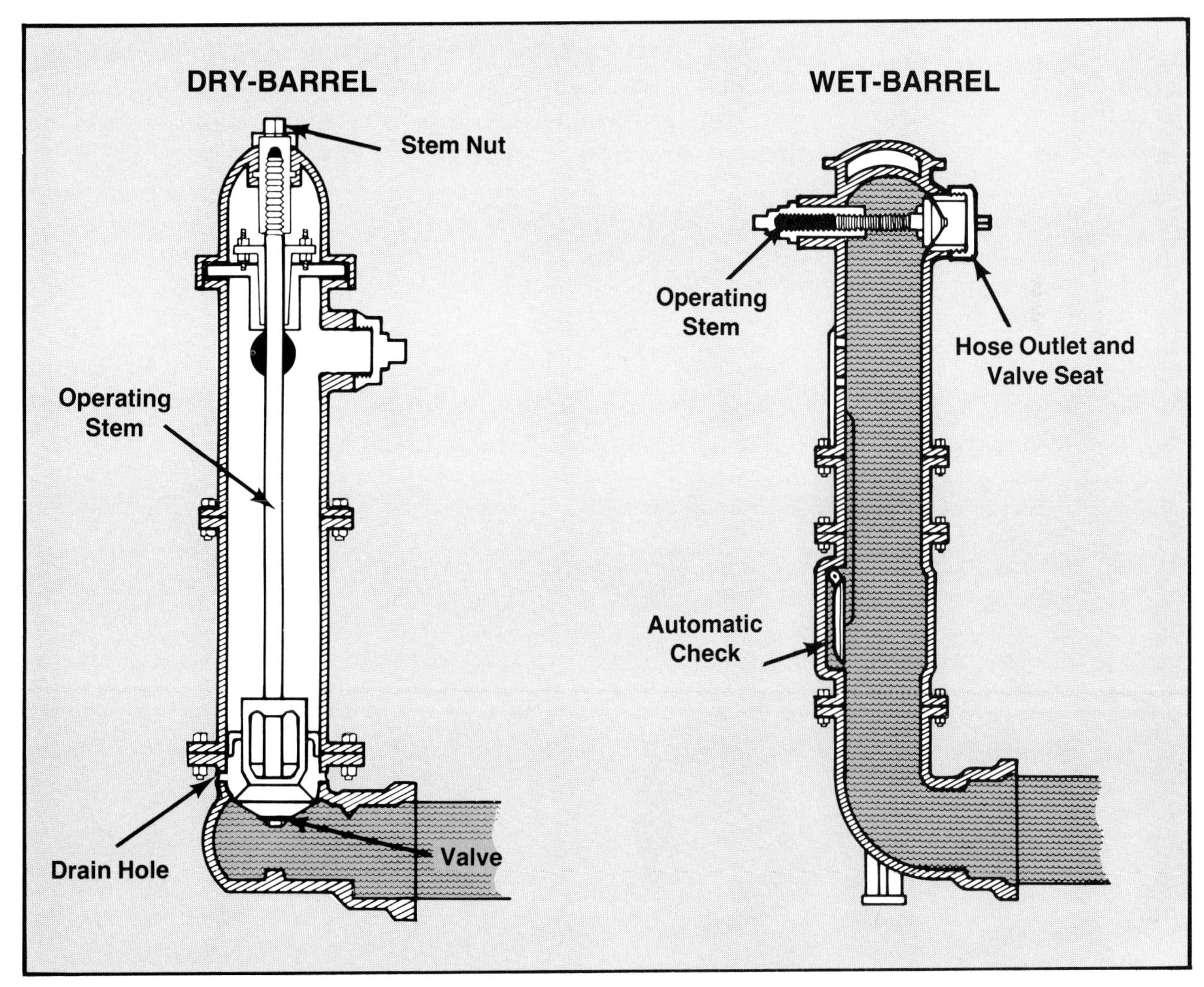

Figure 3.4 Dry-barrel hydrants are used in freezing climates and have valves below the freeze line. Wet-barrel hydrants are used in mild climates and have the valves at the outlets.

valve facings may be made of rubber, leather, or a composition material.

Either type of hydrant is considered standard when equipped with one large outlet (4 or 4½ inch) (100 mm or 115 mm) for pumper connections and two outlets for 2½-inch (65 mm) hose. Hydrant specifications require a 5-inch (125 mm) valve opening for standard 3-way hydrants, and a 6-inch (150 mm) nominal diameter, 120 percent of the valve opening for the waterway and foot piece. All hydrant outlet threads must conform to the hose thread used by the local fire department. National Standard hose coupling threads are desirable for mutual aid operations. Adapters may be necessary when utilizing hydrants in another district or those on a private system.

Current dry-barrel fire hydrant designs generally incorporate a traffic safety flange and operating rod installed just above grade. If a vehicle hits the hydrant, it will shear the hydrant and the operating rod at the flange connection (Figure 3.5). The main valve at the base will remain closed. The blow will not influence the water main integrity. The safety flange also allows a new hydrant and rod to be installed without digging down to the waterway. This provides for less expensive repair and decreases the time out of service.

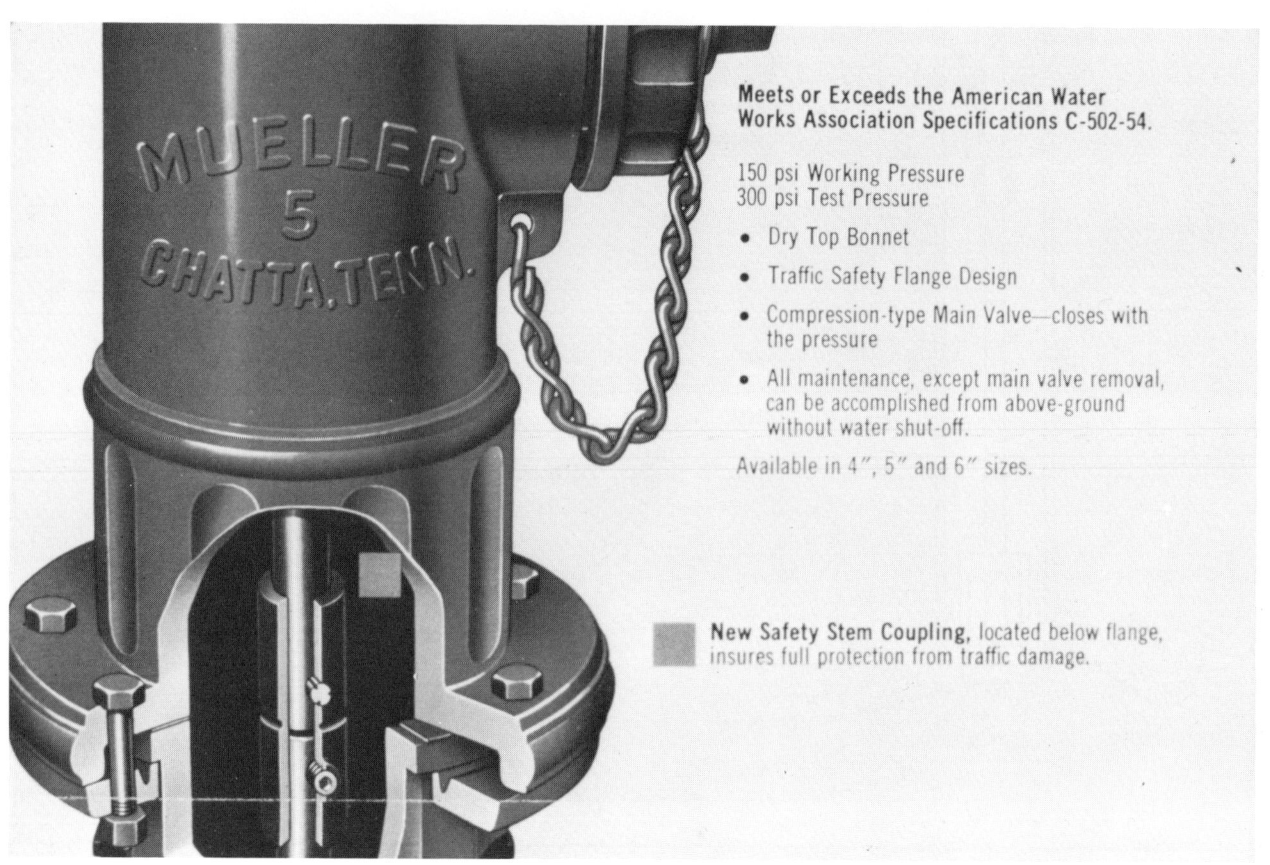

Figure 3.5 Cutaway showing the location of the traffic safety flange. *Courtesy of the Mueller Co.*

HYDRANT INSTALLATION

Water Supply Officer Considerations

The actual installation of fire hydrants should be supervised by the fire department water supply officer or an appointee who understands the fire protection requirements. Standards regarding proper installation can be obtained from the *AWWA #17 Installation, Operation, and Maintenance of Fire Hydrants*. General guidelines are listed here.

The barrel of a fire hydrant should be set plumb to the ground with the lowest discharge outlet at 15 inches (381 mm) above grade. If the hydrant is equipped with a pumper connection (Figure 3.6), it is important to provide sufficient clearance between this connection and the ground. This clearance is necessary to operate the hydrant wrench and the long lugs on the hose coupling. Large hydrant outlets should face the street or road and toward the pumper. Where hydrants are installed before grading is completed, the final grade line should be considered. The distance the hydrant is set from the curb line should not be excessive to enhance direct hydrant connections and large diameter hose operations. Fire hydrants should not be installed close to trees, posts, signs, mail boxes, fences, or other obstructions that impair fire department operations (Figure 3.7).

Figure 3.7 The proximity of this sign pole may obstruct fire fighting operations by making it impossible to connect to the hydrant.

Figure 3.6 If hydrants have a 4- or 4½-inch (100 mm or 115 mm) connection, pumpers can hook up to them with large diameter hose for maximum water supply operations.

Figure 3.8 The gravel at the base of this dry-barrel hydrant facilitates proper drainage of the hydrant.

Drainage should be provided for dry-barrel hydrants by excavating a pit 2 or 3 feet (0.6 m or 0.9 m) below the base of the hydrant. The pit should then be completely filled with coarse gravel or broken stone (Figure 3.8).This material should also be placed around the bowl of the hydrant to a level 6 inches (150 mm) above the drain opening in order to permit proper draining of the hydrant. To facilitate hydrant repair without interruption of the water supply, a gate valve should be installed between the water main and the fire hydrant (Figure 3.9).

The size and shape of the operating nut as well as the direction in which the hydrant stem operates should be uniform for all hydrants in the distribution system. Nonuniformity tends to encourage the use of a pipe wrench instead of a standard hydrant wrench. Pipe wrenches cut the corners of the operating nuts and render them inoperable.

Figure 3.9 The gate valve is installed between the hydrant and the water main. Part of it may extend through the ground to ease in location of the valve.

Color Coding

Fire hydrants should be appropriately marked regarding their single flow as determined by fire flow tests. The recommended marking in NFPA Standard 291, *Fire Flow Testing and Marking of Hydrants* is as follows:

Class AA — 1,500 gpm or greater (5 678 L/min or greater), Light Blue

Class A — 1,000-1,499 gpm (3 785 L/ min to 5 674 L/min), Green

Class B — 500-999 gpm (1 893 L/min to 3 782 L/min), Orange

Class C — Less than 500 gpm (less than 1 893 L/min), Red

Only the caps and bonnets should be color coded with the barrel painted a different distinctive color. Care should be taken to see that paint does not get on the threads in the cap or on the hydrant. Dried paint on the threads could lock the caps in place.

Location Markers

A wide variety of methods are used for fire hydrant identification. Fluorescent paint can be used to aid rapid distinction. Signs similar to roadway regulatory signs are used for spotting hydrant locations along freeways (Figure 3.10). A spotting method frequently used is that of reflective street markers. These markers are placed in the road centerline perpendicular to the fire hydrant. They should be a distinctive color and separate from ordinary lane markers (Figure 3.11). In areas of deep snowfall, metal flags are often attached to hydrants to indicate their location in the drifts or plowed snow (Figure 3.12).

Figure 3.10 Hydrants along freeways may be difficult to spot and should be marked with a sign.

Figure 3.12 Flags may be added to hydrants in areas subject to heavy snowfalls. This will aid in locating the hydrant.

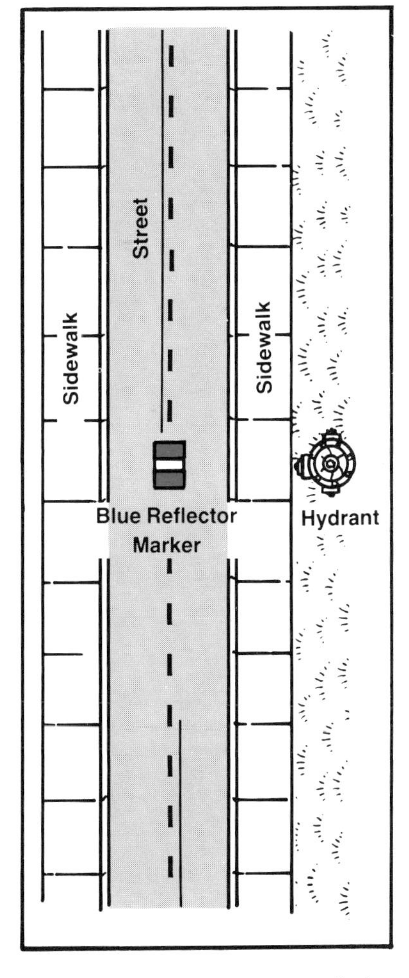

Figure 3.11 Blue reflectors located in the center of a street, directly in front of a hydrant, help the apparatus driver/operator locate the hydrant, particularly at night.

HYDRANT DISTRIBUTION

Ideally, the process of locating fire hydrants in new areas and recommending additional ones in built-up areas should be the responsibility of the fire department water supply officer. However, plan review is generally conducted in the fire prevention bureau or building department, with hydrants located conveniently at intersections. Fire and water department officials should cooperate to develop a policy on hydrant distribution and a procedure for the approval of planned hydrant locations by all concerned agencies. The number of hydrants required should be based on two factors: an adequate number to deliver the required fire flow, and spacing to provide this flow without excessively long hose lays. Fire hydrants must be located and spaced according to the demands of each particular location.

Proper planning will identify development areas where the water supply and hydrant system will require change. Initial installation of an adequate system of hydrants is much easier than costly retrofits.

In high value or high life hazard districts, one or two hydrants should be located at each street intersection with intermediate hydrants so located that spacing does not exceed 300 feet (91 m). In residential districts, one hydrant should be located at each street intersection with intermediate hydrants so located that spacing does not exceed 500 feet (152 m). They should be located adjacent to paved roadways suitable for fire apparatus and, where possible, spaced at least 50 feet (15 m) from any building.

To meet insurance and regulating agency specifications, fire protection engineers or Insurance Services Office Commercial Risk Services, Inc. should be consulted. ISO determines standard hydrant distribution by the average area per hydrant for each fire flow required (Figure 3.13).

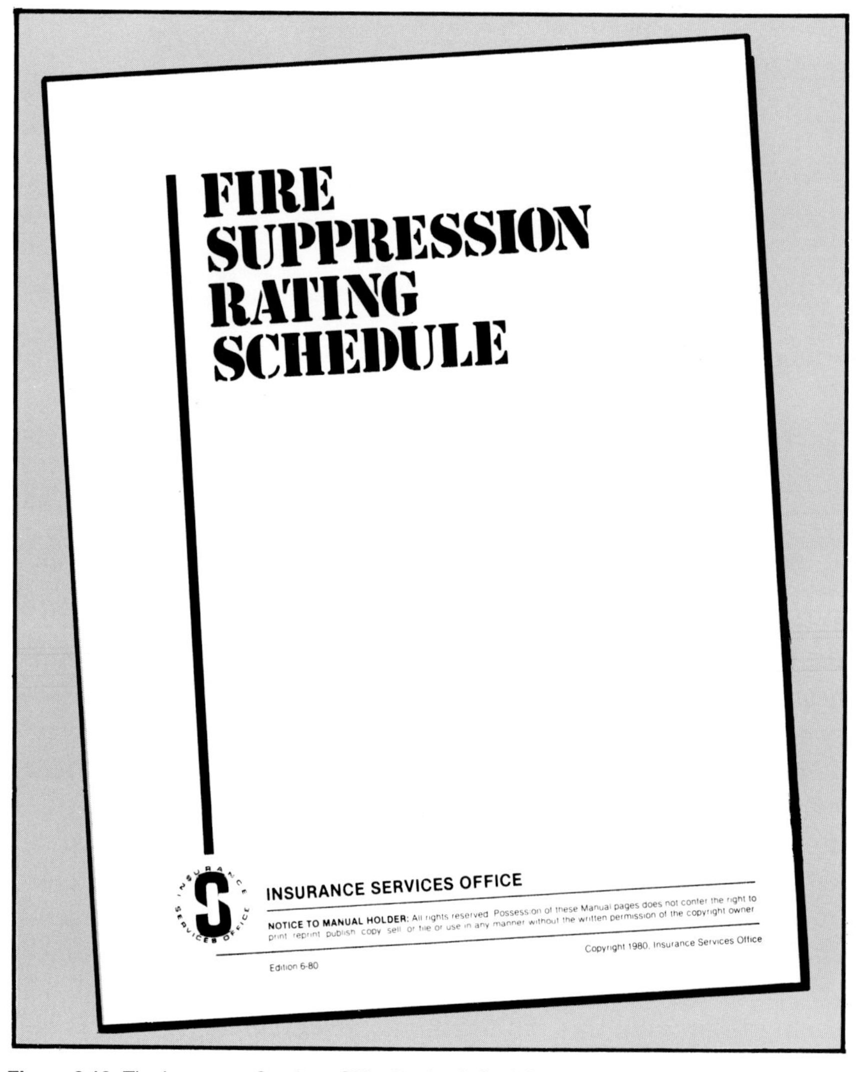
FIRE SUPPRESSION RATING SCHEDULE

INSURANCE SERVICES OFFICE

NOTICE TO MANUAL HOLDER: All rights reserved. Possession of these Manual pages does not confer the right to print, reprint, publish, copy, sell, or file or use in any manner without the written permission of the copyright owner.

Copyright 1980, Insurance Services Office

Edition 6-80

Figure 3.13 The Insurance Services Office Rating Schedule serves as a standard for local fire protection.

HYDRANT INSPECTION

In most cities, repair and maintenance of fire hydrants are the responsibility of the water department. A fire department representative should make inspections to determine if fire hydrants are in good operating condition, and report to the water department all hydrants in need of repair. Fire hydrants should be inspected twice each year, preferably in the spring and fall as well as after each use. A large percentage of faulty hydrant operation may be caused by the misuse of hydrants by public works employees and other persons who are sometimes permitted to operate hydrants. They frequently do not realize the importance of hydrant care, and this misuse may result in a serious fire loss.

In order to make a complete inspection of fire hydrants, an inspector should be equipped with the following:

- loose-leaf notebook
- gauging device for checking threads
- mix of light lubricating oil and graphite
- small flat brush
- gate valve key
- pressure gauge and tapped hydrant cap
- hand pump for pressure up to 200 psi (1 379 kPa)
- 12-quart (11 L) pail
- hydrant wrench
- hydrant record forms
- hydrant repair reports

Procedure

Rust and other sediments are present in water mains and when large flows are made for hydrant inspections, these deposits will be disturbed. To help eliminate complaints of rusty water by water users, the following sequence should be used. On each large main, begin with the hydrant nearest the source of supply. Flow that hydrant until the water clears. Then go to the next hydrant on the main and repeat. Work down that main until a point where it joins another large main is reached. Start on that large main, beginning again at the source. This procedure will greatly reduce complaints.

Water department officials should be notified when hydrant inspections are to be made, and they should be kept informed concerning the location and route to be taken. The following conditions should be checked at each hydrant:

- Check for any obstructions that have been erected near the hydrant. These obstructions generally include traffic standards, protective barriers, sign posts, utility poles, shrubbery, and fences (Figure 3.14).

- Check to see if the hydrant outlets face the proper direction and if there is sufficient clearance between the hydrant outlet and the surrounding ground. The clearance between the bottom of the butt and the grade should be at least 15 inches (381 mm) (Figure 3.15).
- Check to see if the hydrant has been damaged by traffic or if it is set too close to the curb exposing it to vehicle traffic (Figure 3.16).
- Check the condition of the paint.

Figure 3.14 Excess brush and shrubbery surrounding a hydrant should be removed so that the hydrant is clearly visible.

Figure 3.15 It should not be necessary to use a shovel in order to make a hydrant connection. Proper installation of hydrants can prevent this situation.

Figure 3.16 Damage to hydrants caused by vehicles can be prevented by installing barriers around hydrants. This is particularly useful where the potential for such damage is high, such as in parking lots.

Before opening any fire hydrant, the inspector should check for foreign material inside the hydrant outlet. The presence of water in a dry-barrel hydrant indicates that either the main valve does not hold tightly or the drain holes are plugged. The location of the gate valve in the hydrant branch should be known so that it can be closed if complications are encountered after the hydrant is opened. The hydrant valve should be fully opened and water permitted to flow until clear to flush the hydrant branch and hydrant. After flushing, close the hydrant slowly to prevent water hammer in the main.

After the hydrant is closed, it should be subjected to a pressure test. A pressure test requires a tapped hydrant cap with gauge attached to one outlet. The caps on the other outlets should be tightened, and the hydrant valve should then be completely opened. The static pressure should be recorded, and the tightness of the nipples and stuffing box observed (Figure 3.17 on next

Figure 3.17 Before recording the static pressure, the air should be bled from the hydrant until a steady stream flows through the petcock nozzle.

page). The hydrant should then be closed and observations should be made to see if it drains properly. If the hydrant does not drain freely, the pressure pump should be connected to the hydrant by using the tapped hydrant cap. Water should be pumped from the pail into the hydrant barrel to a pressure of 200 psi (1 379 kPa) if necessary. This pressure will force most obstructions from around the drain valve, and in most cases, provide a simple remedy for faulty drainage. If a hand pump is not available, special connections for a booster hoseline from a fire department pumper may be used. In either case, the hydrant valve must be completely closed.

If there is any considerable leakage at the stuffing box, the main stem should be repacked and oiled. If the leakage is slight, tightening the stuffing box nut will probably stop the leak. In any event, before leaving the hydrant, the packing in the stuffing box should be oiled since packing tends to dry over a period of time. The valve stem should be oiled where provision for oiling is made (Figure 3.18). A thread gauge device should be screwed on each outlet to determine if a proper fit can be made. The threads in each of the hydrant caps and on the outlets should be lightly swabbed with a mixture of lubricating oil and graphite (Figure 3.19). The threads should then be wiped clean with a rag. If there is rust or deposits inside the cap or on the outlet, it should be removed with a wire brush before lubricant is applied (Figure 3.20). Gaskets should be replaced if necessary, and the caps should be placed on the discharge outlet and closed firmly (Figure 3.21). Chains, if provided, should be freed of excess paint and straightened to ensure free running around the groove in the cap. This procedure

Figure 3.18 Oiling the valve stem makes it easier to open and close the hydrant.

Figure 3.19 Oiling the discharge orifice threads and the hydrant cap threads will make cap removal and hose connections much easier.

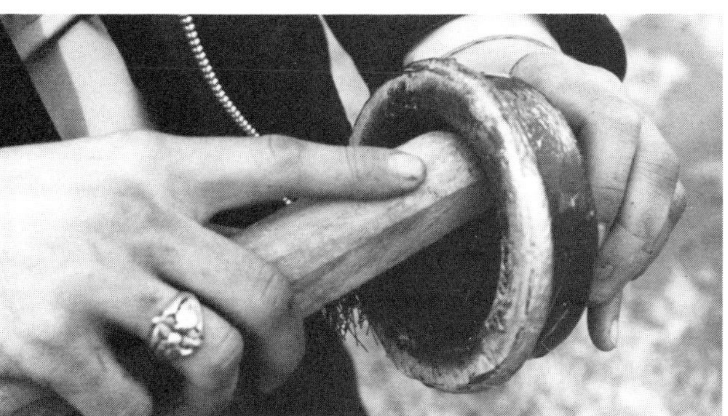

Figure 3.20 Remove any debris in the hose threads with a wire brush.

Figure 3.21 Inspect the cap gasket for damage and replace it if necessary.

should leave the hydrant in good condition. Regular inspections will help ensure good working conditions of all hydrants and minimize later fireground problems.

The fire department should maintain complete records of the inspection results for each hydrant. A sample inspection sheet is shown in Figure 3.22. All hydrant records should be kept for five years. Any needed repairs should be reported in writing to the water department representative. Follow-up inspections should be conducted and recorded to determine if the needed repairs have been promptly and properly made.

HYDRANT RECORD

LOCATION______ HYDRANT NO.______
POSITION______ MAKE______
INSTALLED______ TYPE______ TURNS TO OPEN____ R.____ L.______
SIZE OF LEAD______ SIZE OF MAIN______
VALVE IN LEAD______ FT.______ TURNS TO OPEN____ R.____ L.______
BENCH MARK______ ELEV.______

PRESSURE TESTS

DATE	STATIC PRESSURE	FLOW PRESSURE	GPM	DATE	STATIC PRESSURE	FLOW PRESSURE	GPM

REMARKS

RECORD OF MAINTENANCE

WORK PERFORMED______ DATE______

Flowed												
Lubricated												
Cap Gasket Replaced												
Bonnet Gasket Replaced												
Valve Leather Replaced												
Drain Valve Repaired												
Cap Replaced												
Lead Valve Operated												
Painted												
Raised												
Moved												

Figure 3.22 A hydrant inspection form can be used both for recording field information and as a permanent record. Modern record-keeping systems utilize computers.

Many communities are promoting "Adopt A Hydrant" or similar programs in which civic groups or individuals conduct the small scale maintenance items that make hydrants easier to use. These maintenance items include clearing hydrants of weeds and debris, shoveling snow from around them, not planting by hydrants, and other similar concerns.

MAPS

Complete records can be valuable aids to ensure the proper installation and maintenance of a water supply and distribution system. A display of up-to-date, accurate, and legible maps of a water distribution system is very important when pre-incident plans are being made. Copies of water supply maps may be made from the water utility's master copy. Corrections and extensions should be added as a community and its water system expand. In this way, a water distribution map will not become outdated and useless.

Each fire station should have a map that locates the following:

- Water Mains. It is important for maps to show the main to which the hydrant is connected, because one hydrant in an area may deliver an ample supply while another will not. In large cities, it is best to have one map that shows 12-inch (305 mm) and larger arterial mains for the overall system and a second map that shows all mains and hydrants in that part of the city in which a fire company operates.
- Hydrants. All hydrants should be located accurately on the map, including their proper location at street intersection corners. Fire flow test results should be studied and hydrants with inadequate flows for pumper supply should be accordingly noted.

All maps should be mounted on a fire station wall and studied by the firefighters and officers (Figure 3.23). Maps should also be

Figure 3.23 Water supply maps should be available to dispatch personnel so they can relay information to responding fire units. Modern information systems use computers to provide this information.

provided at communication centers for reference by field units. Photocopy reductions should be placed in command units and individual pumpers (Figure 3.24).

Figure 3.24 Books containing maps showing the locations of hydrants and other water supply sources can be carried on the apparatus to assist responding crews.

A water distribution map should be used to fulfill the needs of the local fire department. Various colors should be used to designate the different pipe sizes in the system. A suggested color scheme that could be used is the following:

- 4-inch (100 mm) pipe — red
- 6-inch (150 mm) pipe — yellow
- 8-inch (200 mm) pipe — brown
- 10-inch (250 mm) pipe — blue
- 12-inch (305 mm) or greater — green

The colors used are not significant as long as a legend on the map identifies the main size with its corresponding color. Fire flows within given areas or for certain buildings may also be included on the map. A typical section map is shown in Figure 3.25.

Many municipalities are now utilizing computers to maintain hydrant and water main records and information. Particularly in larger areas, the computer is the quickest method of obtaining information on hydrants and water mains in any specific location. This can be extremely helpful information to crews responding to an emergency incident.

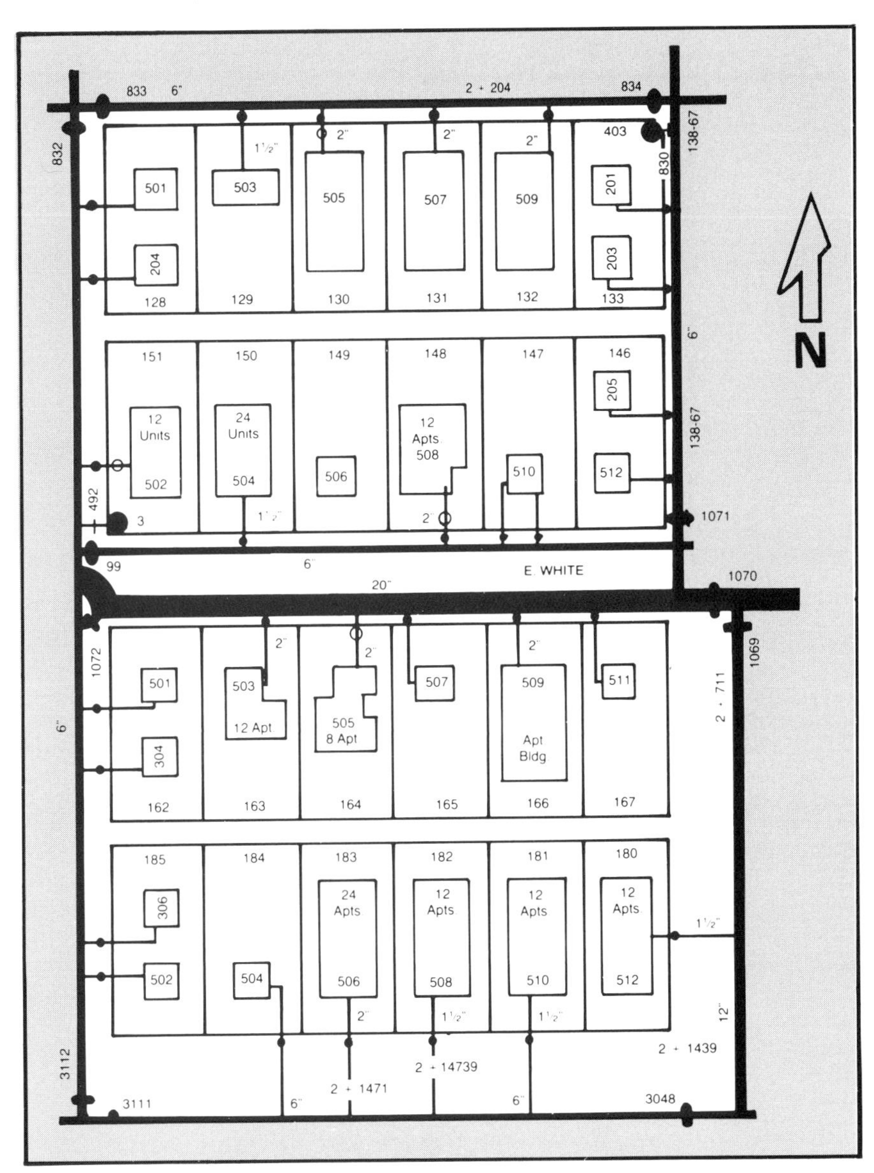

Figure 3.25 Water departments can usually supply maps of water supply systems.

Chapter 3 Review

Answers on page 260

TRUE-FALSE. Mark each statement true or false. If false, explain why.

1. A wet-barrel hydrant is more likley to be found in colder climates.

   ☐ T ☐ F ______________________________

   ____

2. Fire hydrants should be inspected twice a year and after each use.

   ☐ T ☐ F ______________________________

   ____

3. During hydrant testing, it is best to save water by closing the hydrant quickly once the color of the water clears.

   ☐ T ☐ F ______________________________

   ____

4. Hydrant inspection records should be kept for two years.

   ☐ T ☐ F ______________________________

   ____

5. Pipe wrenches may be used routinely in place of hydrant wrenches.

   ☐ T ☐ F ______________________________

   ____

MULTIPLE CHOICE: Circle the correct answer.

6. Which of the following discharge outlet configurations would qualify a hydrant to be considered a standard fire hydrant?
   A. One 5-inch (125 mm) and two 2½-inch (65 mm) outlets
   B. One 6-inch (150 mm) and two 2½-inch (65 mm) outlets
   C. Two 2½-inch (65 mm) outlets
   D. One 4- or 4½-inch (100 mm or 115 mm) and two 2½-inch (65 mm) outlets

7. A hydrant discharge should be at least ________ inches (mm) from the ground.
   A. 12 inches (305 mm)
   B. 15 inches (380 mm)
   C. 18 inches (457 mm)
   D. 24 inches (610 mm)

8. When planning a water supply system for an area, it is important to consider the number and ________ of hydrants.
   A. spacing
   B. size
   C. color
   D. rating

MATCHING: Write the correct letter in the space provided.

9. Match the hydrant flow rating to the appropriate color.

_______	Green	A. 1,500 gpm (5 678 L/min) or greater
_______	Red	B. 1,000 to 1,499 gpm (3 785 L/min to 5 674 L/min)
_______	Light Blue	C. 500 to 999 gpm (1 893 L/min to 3 782 L/min)
_______	Orange	D. Less than 500 gpm (1 893 L/min)

SHORT ANSWER: Answer each item briefly.

10. Name the two types of fire hydrants and briefly describe when each is used.

A. ____

B. ____

11. A residential development is planned in your district. The plans show a hydrant at every intersection and hydrants spaced every 750 feet (229 m) apart between intersections. Is this a suitable arrangement?

____

____

12. Name at least three conditions that should be checked at each hydrant when testing and inspecting.

____

____

____

13. Who is responsible for the spacing of fire hydrants?

____

____

____

Chapter 4

Fire Flow Testing

This chapter provides information that addresses performance objectives in NFPA 1001, *Fire Fighter Professional Qualifications* (1987) and NFPA 1002, *Fire Apparatus Driver/Operator Professional Qualifications* (1988), particularly those referenced in the following sections:

NFPA 1001

Fire Fighter III

4-15.2

4-15.3

4-15.4

NFPA 1002

3-2.3

3-2.4

3-2.5

Chapter 4

Fire Flow Testing

In order for fire service personnel to determine the quantity of water available for fire protection, it is necessary to conduct fire flow tests on the water distribution system. These tests include the actual measurement of static (normal operating) and residual pressures, and flow from hydrants. This section describes the test procedure, equipment needed, and the formulas and calculations used to determine available water.

REASONS FOR TESTING

Fire flow tests are made to determine the rate of water flow available for fire fighting at various locations within the distribution system. By measuring the flow from hydrants and recording the pressures corresponding to this flow, the number of gallons (liters) available at any pressure or the pressure available at any flow can be determined through calculations or graphical analysis.

Before conducting a flow test, a responsible water department official should be notified since the opening of hydrants may upset the normal operating conditions in a water supply system. Also, water service personnel may be doing maintenance work in the immediate vicinity and, therefore, the results of the test would not be typical for normal conditions. This practice of proper notification will also promote a better working relationship between waterworks and fire service personnel.

Knowing the capacity of a water system is just as important as knowing the capacities of pumpers and water tanks. This knowledge is also essential when making pre-incident plans. The results of fire flow tests can be used to an advantage by both the fire and water departments of a municipality. Fire officers who are familiar with fire flow test results are better qualified to locate pumpers at strong locations on a distribution system while avoiding weak locations. Since test results indicate weak points in a water distribution system, they can be used by water works personnel when improvements in an existing system are planned, and when extensions to newly developed areas are designed. Tests that are repeated at the same locations year after year may reveal a loss in the carrying capacity of water mains and a need for strengthening certain arterial mains. Flow tests should be run after any extensive water main improvements, after extensions have been made, or at least every five years if there have been no changes.

It is important for every fire officer to know the ultimate capacity of the water system. Fire fighting defenses cannot be intelligently planned without all the known facts. The overall capacity of the water system and its flow in given areas are certainly important factors. A hydrant flow test for an area is a means by which certain facts can be established. Flow tests are the only positive means to determine the quantity of water available for fire fighting.

USING THE PITOT TUBE AND GAUGE

The use of a pitot tube and gauge for taking a flow reading is not difficult, but it is essential that it is used properly if accurate readings are to be obtained. Two designs of a pitot tube and gauge are shown in Figure 4.1. The handle serves as an air chamber to help keep the gauge needle steady. It is possible for a department to construct its own pitot tube (Figure 4.2).

A good method of holding a pitot tube and gauge in relation to a hydrant outlet or nozzle is to grasp the pitot tube just behind the blade with the first two fingers and thumb of the left hand while the right hand holds the air chamber. The little finger of the left hand rests upon the hydrant outlet or nozzle tip to steady the instrument (Figure 4.3).

Unless some effort is made to steady the pitot tube, the movement of the water will make it difficult to get an accurate reading.

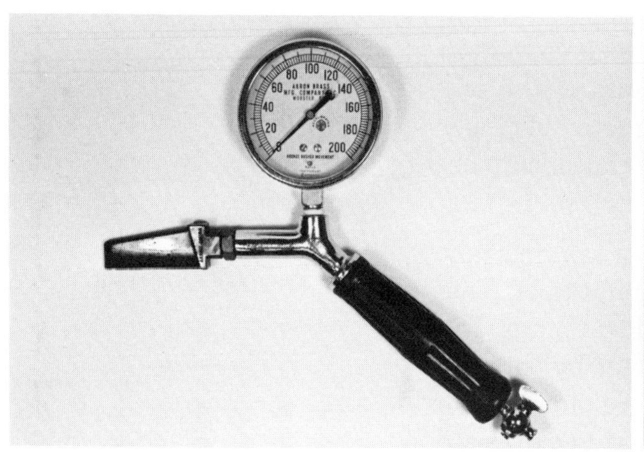

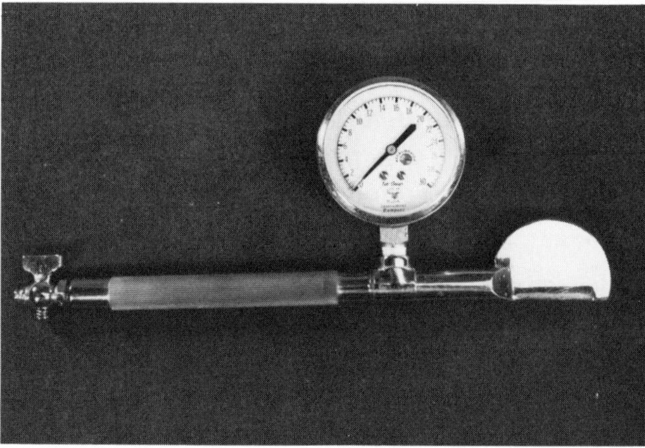

Figure 4.1 Pitot tubes come in several different designs, but they all operate the same way.

Figure 4.2 Given the necessary components, it is possible to construct a homemade pitot tube. Two different homemade versions are shown here.

Figure 4.3 The little finger is used to help steady the pitot tube.

Another method of holding the pitot tube is illustrated in Figure 4.4. The left hand fingers are split around the gauge outlet and the left side of the fist is placed on the edge of the hydrant orifice or outlet. The blade can then be sliced into the stream in a coun-

Figure 4.4 Steady the pitot by holding the left side of the fist against the discharge outlet. Then slice the blade into the stream.

terclockwise direction. The right hand once again steadies the air chamber. The procedure for using a pitot tube and gauge is as follows:

- Open the petcock on the pitot tube and make certain the air chamber is drained, then close the petcock.
- The blade is edged into the stream, with the small opening or point being centered in the stream and held away from the butt or nozzle approximately one-half the diameter of the opening (Figure 4.5). For a 2½ inch (65 mm) hydrant butt, this distance would be 1¼ inches (32 mm).
- The pitot tube blade should now be parallel to the outlet opening with the air chamber kept above the horizontal plane passing through the center of the stream. This will increase the efficiency of the air chamber and help avoid needle fluctuations.
- Take and record the velocity pressure reading from the gauge. If the needle is fluctuating, read and record the value located in the center between the high and low extremes.

After the test is completed, open the petcock and be certain all water is drained from the assembly before storing.

Figure 4.5 When the pitot is inserted into the stream, it should be located directly in the middle of the stream at a distance from the orifice equal to one-half the diameter of the orifice.

COMPUTING HYDRANT FLOW

The easiest way to determine how much water is flowing from the hydrant outlet(s) is to refer to prepared tables for nozzle discharge. These tables have been computed by using a formula for gallons per minute (L/min) flow when the flow pressure is known. The formula may be stated as follows: Flow rate is equal to a constant multiplied by the coefficient of discharge, multiplied by the diameter of the orifice squared, multiplied by the square root of the pressure. The formula is written as follows:

$$GPM = 29.83 \times C \times d^2 \times \sqrt{P}$$

$$L/min = 0.0667766 \times C \times d^2 \times \sqrt{P}$$

where

d is the actual diameter of the hydrant or nozzle orifice in inches (mm).

P is the pressure in psi (kPa) as read on the gauge of the pitot at the orifice.

C is the coefficient of discharge.

NOTE: 29.83 (0.0667766) is a constant derived from the physical laws relating water velocity, pressure, and conversion factors that conveniently leave the answer in gallons per minute (liters per minute). The metric formula is not a direct conversion of the U.S. formula.

This formula was derived by assigning a coefficient of 1.0 for an ideal frictionless discharge orifice. An actual hydrant orifice or nozzle will have a lower coefficient of discharge, reflecting friction factors that will slow the velocity of flow. The coefficient will vary with the type of hydrant outlet or nozzle used. When using a hydrant orifice, the operator will have to feel the inside contour of the hydrant orifice to determine which one of the three types is being used (Figure 4.6). When a nozzle is used, the coefficient of discharge will depend on the type. Some common nozzle types and their respective coefficients of friction are given in (Figure 4.7).

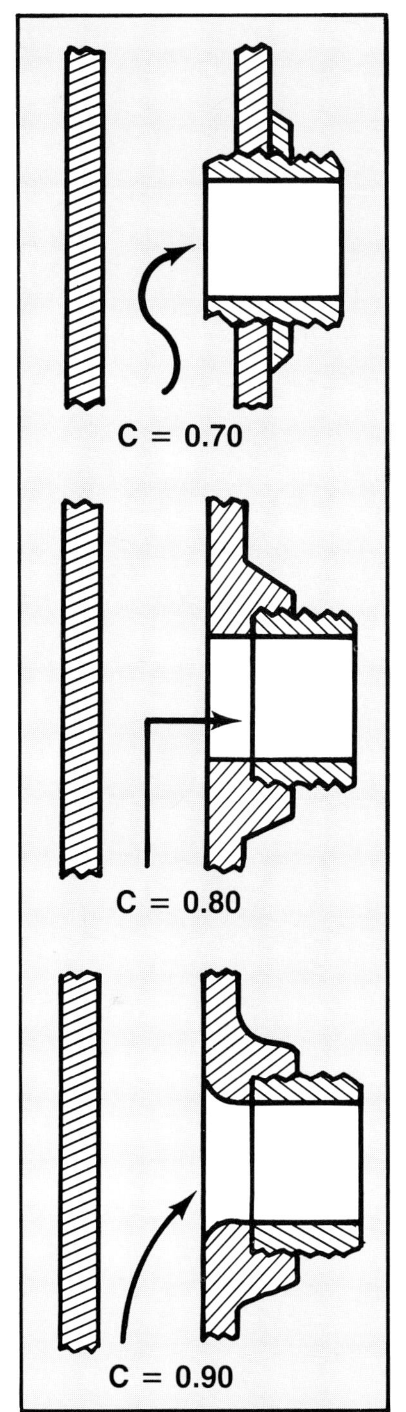

Figure 4.6 Shown are three types of hydrant discharges and their respective discharge coefficients.

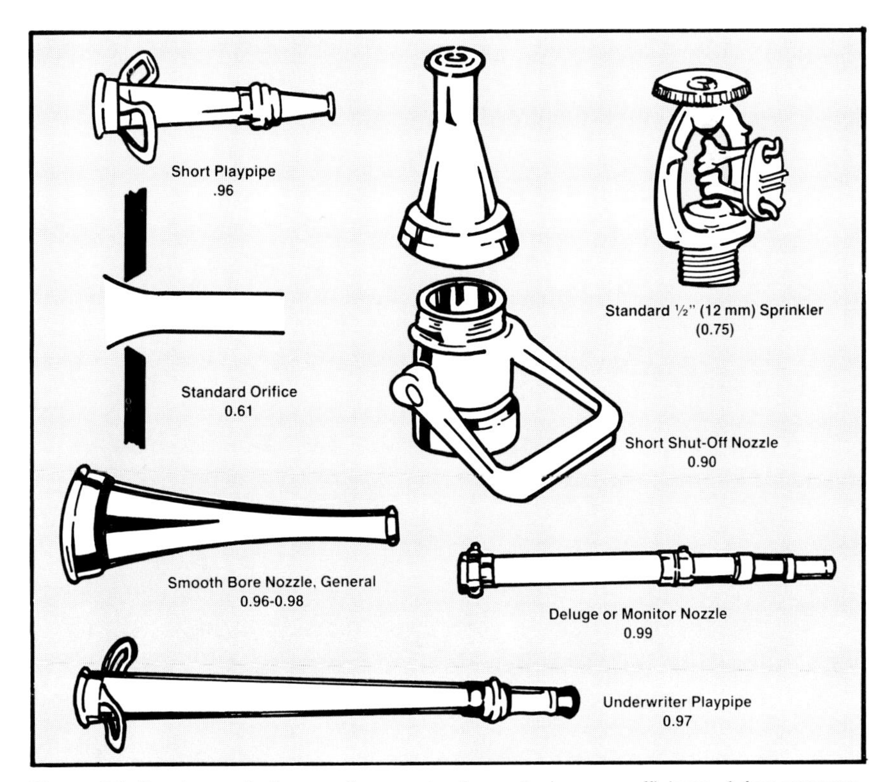

Figure 4.7 Nozzles and other appliances also have discharge coefficients. A few common ones are shown here.

This flow formula also depends on the actual internal diameter of the outlet or nozzle opening being used. A ruler with a scale that measures to at least sixteenths of an inch (mm's) should be used to measure the diameter of the outlet or nozzle opening (Figure 4.8).

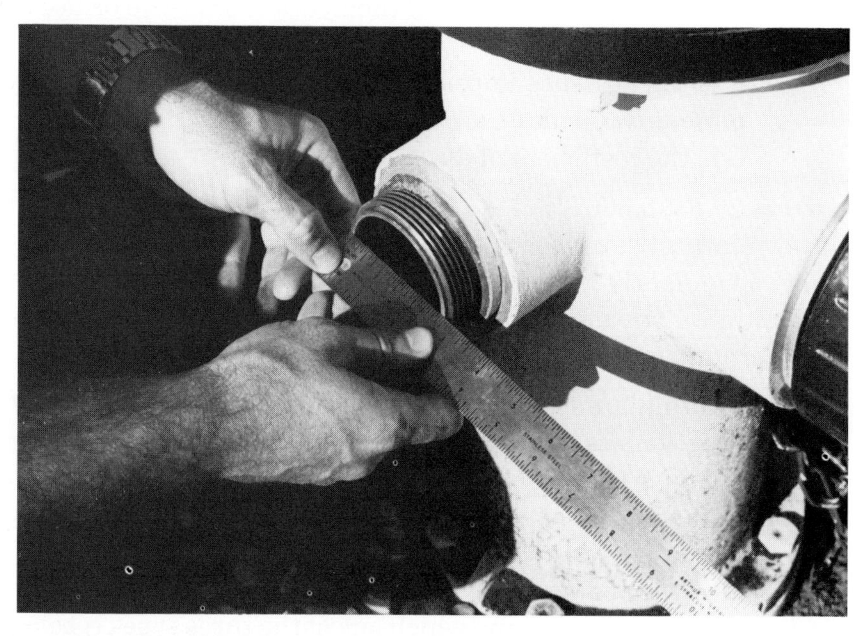

Figure 4.8 Use a ruler to measure the exact diameter of the hydrant discharge orifice; the one shown above is actually about 2⅝ inches (67 mm).

Assuming a 2½-inch (65 mm) hydrant outlet is used that has an actual diameter of 27/16 inches or 2.44 inches (62 mm) with a C factor of 0.80, and a flow pressure of 10 psi (69 kPa) read from the pitot gauge, the waterflow equation would read:

$$\text{GPM} = 29.83 \times C \times d^2 \times \sqrt{P}$$
$$= 29.83 \times 0.80 \times (2.44)^2 \times \sqrt{10}$$
$$\text{GPM} = 449.28 \text{ or } \approx 450$$
$$\text{L/min} = 0.0667766 \times C \times d^2 \times \sqrt{P}$$
$$= 0.0667766 \times 0.80 \times (62)^2 \times \sqrt{69}$$
$$\text{L/min} = 1705.78 \text{ or } 1700$$

Table 4.1 (U.S. and metric) gives flows for different outlet sizes and pitot gauge readings.

The stream from a large hydrant outlet (4 to 4½ inches) (100 mm to 115 mm) contains voids (i.e., the entire stream of water is not solid), and for this reason the above formula alone will not give accurate results for flows using these large outlets. Generally, the 2½ inch (65 mm) outlets should be used. If it is necessary to use the large outlets, a correction factor can be used to give fairly accurate results. These factors are given in Table 4.2. The flow (as determined by $\text{GPM} = 29.83 \times C \times d^2 \times \sqrt{P}$ or $\text{L/min} = 0.0667766 \times C \times d^2 \times \sqrt{P}$) should be multiplied by one of the factors corresponding to the velocity pressure measured by the pitot tube and gauge.

TABLE 4.1
DISCHARGE TABLE FOR CIRCULAR OUTLETS* (U.S.)
Outlet Pressure Measured by Pitot Gauge

Outlet Pressure in lbs. per sq. inch	OUTLET DIAMETER IN INCHES											
	2 3/8	2 1/2	2 5/8	2 3/4	2 7/8	3	3 1/8	3 7/8	4	4 3/8	4 1/2	4 5/8
	U.S. Gallons per Minute											
1	150	170	180	200	220	240	260	400	430	510	540	580
2	210	240	260	290	310	340	370	570	610	720	770	810
3	260	290	320	350	380	420	450	700	740	890	940	990
4	300	340	370	410	440	480	530	810	860	1030	1090	1150
5	340	380	410	450	500	540	590	900	960	1150	1220	1290
6	370	410	450	500	540	590	640	990	1050	1260	1340	1410
7	400	440	490	540	590	640	690	1070	1140	1360	1440	1520
8	430	480	520	570	630	680	740	1140	1220	1450	1540	1620
9	450	500	550	610	670	730	790	1210	1290	1540	1640	1720
10	480	530	580	640	700	760	830	1280	1360	1630	1730	1820
11	500	560	610	670	730	800	870	1340	1430	1710	1810	1910
12	520	580	640	700	770	840	910	1400	1490	1780	1890	1990
13	550	610	670	730	800	870	950	1450	1550	1850	1960	2070
14	570	630	690	760	830	900	980	1510	1610	1920	2040	2150
15	590	650	720	790	860	940	1020	1560	1660	1990	2110	2220
16	610	670	740	810	890	970	1050	1620	1720	2060	2180	2300
17	620	690	760	840	910	1000	1080	1660	1770	2120	2240	2370
18	640	710	780	860	940	1030	1110	1710	1820	2180	2310	2440
19	660	730	810	890	960	1050	1140	1760	1870	2240	2370	2510
20	680	750	830	910	990	1080	1170	1800	1920	2290	2430	2570
22	710	790	870	950	1040	1130	1230	1890	2020	2400	2550	2700
24	740	820	910	1000	1090	1180	1290	1970	2110	2510	2660	2810
26	770	860	940	1040	1130	1230	1340	2050	2190	2620	2770	2930
28	800	890	980	1070	1170	1280	1390	2130	2280	2720	2880	3040
30	830	920	1010	1110	1210	1320	1430	2210	2350	2820	2980	3150
32	860	950	1050	1150	1260	1370	1480	2280	2430	2910	3080	3250
34	880	980	1080	1180	1290	1410	1530	2350	2510	3000	3170	3350
36	910	1010	1110	1220	1330	1450	1580	2420	2580	3080	3260	3440
38	930	1040	1140	1250	1370	1490	1620	2480	2650	3170	3350	3540
40	960	1060	1170	1290	1400	1530	1660	2550	2720	3250	3440	3630

*Computed with Coefficient C = 0.90, to nearest 10 gallons per minute.

TABLE 4.1
DISCHARGE TABLE FOR CIRCULAR OUTLETS* (Metric)
Outlet Pressure Measured by Pitot Gauge

Outlet Pressure in kPa	OUTLET DIAMETER IN MM											
	60	64	67	70	73	76	79	98	102	111	114	117
	Liters per Minute											
5	484	550	603	658	716	776	839	1291	1398	1656	1746	1840
10	684	778	853	931	1012	1098	1186	1825	1977	2341	2470	2602
15	838	953	1045	1140	1240	1344	1453	2235	2422	2868	3025	3186
20	968	1101	1206	1317	1432	1552	1677	2581	2796	3312	3493	3679
25	1082	1231	1348	1472	1601	1736	1875	2886	3126	3702	3905	4113
30	1185	1348	1478	1613	1754	1901	2054	3161	3425	4056	4278	4506
35	1280	1456	1596	1742	1894	2054	2219	3415	3699	4381	4620	4867
40	1368	1557	1706	1862	2026	2195	2372	3650	3954	4683	4940	5203
45	1451	1651	1810	1975	2148	2328	2516	3871	4194	4967	5239	5519
50	1530	1741	1907	2082	2264	2455	2652	4081	4421	5236	5523	5817
55	1604	1826	2001	2184	2375	2574	2781	4281	4637	5492	5792	6101
60	1676	1907	2090	2281	2481	2689	2905	4471	4843	5736	6050	6373
65	1744	1985	2175	2374	2582	2799	3024	4653	5041	5970	6293	6633
70	1810	2060	2257	2463	2679	2904	3138	4829	5231	6195	6535	6883
75	1873	2132	2336	2550	2773	3006	3248	4999	5415	6413	6764	7125
80	1935	2202	2413	2634	2864	3105	3355	5162	5593	6623	6986	7358
85	1994	2270	2487	2715	2952	3200	3458	5321	5765	6827	7201	7589
90	2053	2335	2559	2794	3038	3293	3558	5476	5932	7025	7410	7805
95	2109	2399	2629	2870	3121	3383	3656	5625	6094	7217	7612	8019
100	2164	2462	2698	2945	3203	3471	3751	5772	6253	7405	7810	8227
105	2217	2522	2764	3017	3282	3557	3843	5914	6407	7589	8003	8430
110	2269	2582	2860	3089	3359	3640	3933	6054	6558	7766	8192	8628
115	2320	2640	2893	3158	3434	3722	4022	6190	6705	7940	8376	8822
120	2370	2697	2955	3225	3508	3803	4109	6323	6849	8112	8556	9012
125	2419	2752	3016	3292	3581	3881	4193	6453	6990	8279	8732	9198
130	2467	2807	3076	3358	3652	3958	4277	6581	7129	8443	8905	9380
135	2514	2860	3135	3422	3721	4033	4358	6706	7265	8604	9075	9559
140	2560	2913	3192	3484	3789	4107	4438	6829	7398	8761	9241	9734
145	2605	2964	3249	3546	3856	4180	4516	6950	7529	8917	9405	9907
150	2650	3015	3304	3607	3922	4251	4594	7069	7658	9069	9569	10076

*Computed with Coefficient C = 0.90, to nearest liter.

TABLE 4.2
CORRECTION FACTORS FOR LARGE DIAMETER OUTLETS

VELOCITY PRESSURE	FACTOR
2 psi (13.8 kPa)	0.97
3 psi (20.7 kPa)	0.92
4 psi (27.6 kPa)	0.89
5 psi (34.5 kPa)	0.86
6 psi (41.4 kPa)	0.84
7 psi (48.3 kPa) or over	0.83

From Table 4.1, a flow of 6 psi (41.4 kPa) through a 4-inch (100 mm) outlet is indicated as 1,050 gpm (3 974 L/min). However, tests have indicated that only 84 percent of this quantity is actually flowing due to the void. Accordingly, actual flow is 1,050 x 0.84 = 882 gpm (3 974 x 0.84 = 3 338 L/min).

These formulas allow the computation of total flow from the flowing hydrants when performing an area fire flow test. They also would indicate the flow from the hydrant at the time of the test.

REQUIRED RESIDUAL PRESSURE

As a result of experience and water system analysis, fire protection engineers have established 20 psi (138 kPa) as the minimum required residual pressure when computing the available water for area flow test results. This residual is considered enough reserve pressure to overcome friction loss in a short 6-inch (150 mm) branch, in the hydrant itself, and in the intake hose, as well as allowing a safety factor to compensate for gauge error. Many state health departments require this 20 psi (138 kPa) minimum to prevent the possibility of external water being drawn into the system at main connections. Pressure differentials can result in water main collapse or create cavitation which is the implosion of air pockets drawn into pumps. A more common occurrence is that pumpers working at these low system pressures may be pumping near the water mains' capacity. If a valve on the pumper is shut down too quickly, a water hammer is created and may be transferred to the water main that can damage or break mains or connections.

FLOW TEST PROCEDURES

When testing the available water supply, the number of hydrants to be opened will depend upon an estimate of the flow that may be available in the area; a very strong probable flow requires several hydrants to be opened for a more accurate test. Enough hydrants should be opened to drop the static pressure by at least 10 psi (70 kPa); however, if more accurate results are required, the pressure drop should bring the residual pressure as close as possible to 20 psi (138 kPa). The flow available at 20 psi (138 kPa) can be determined by dropping the residual pressure to exactly 20 psi (138 kPa) or can be determined at any residual pressure by graphical analysis, or by formula calculations.

Another problem that might be encountered is that water mains may contain such low pressures that no flow pressure will register on the pitot gauge. If this is the case, straight stream nozzles with smaller than 2½ inch (65 mm) orifices must be placed on the hydrant outlet to increase the flow velocity to a point where the velocity pressure is measurable. It should be noted that these straight stream nozzles will require an adjustment in the waterflow calculation that must include the smaller diameter and the respective coefficient of friction.

Flow tests are sometimes conducted in areas very close to the base of an elevated water storage tank or standpipe and result in flows that are quite large in gallons per minute (L/min). It should be realized that such large flows can only be sustained as long as there is sufficient water in the elevated tank or standpipe. It is advisable to make an additional flow test with the storage shut off. The flow obtained from this second test is the quantity available when the storage has been depleted.

The static pressure and the residual pressure during a flow test should be taken from a fire hydrant that is located as close as possible to the location requiring the test results. This hydrant is commonly called the "test" hydrant. The "flow" hydrants are those where pitot readings are taken to find their individual flows, then added to find the total flow during the test.

Five arrangements by which fire hydrants in an area can be selected for a flow test are shown in Figure 4.9. The illustration shows the location of test hydrants relative to the flow hydrants with different hydrant and main configurations. In general, when flowing a single hydrant, the test hydrant should be between the flow hydrant and the water supply source. That is, the flow hydrant should be downstream from the test hydrant. When flowing multiple hydrants, the test hydrant should be centrally located relative to the flow hydrants.

The procedure for conducting an available water test is as follows:

Step 1: Locate personnel at the test hydrant and at all flow hydrants to be used.

Step 2: Remove a hydrant cap from the test hydrant and attach the pressure gauge cap with the petcock in the open posi-

NOTE: The hydrant in the circle is the test hydrant. Residual and static flows are measured here.

Dead End Main — Water One Way

Water From Two Directions

Water From Two Directions At Corner

Water From Three Directions

Water From Four Directions

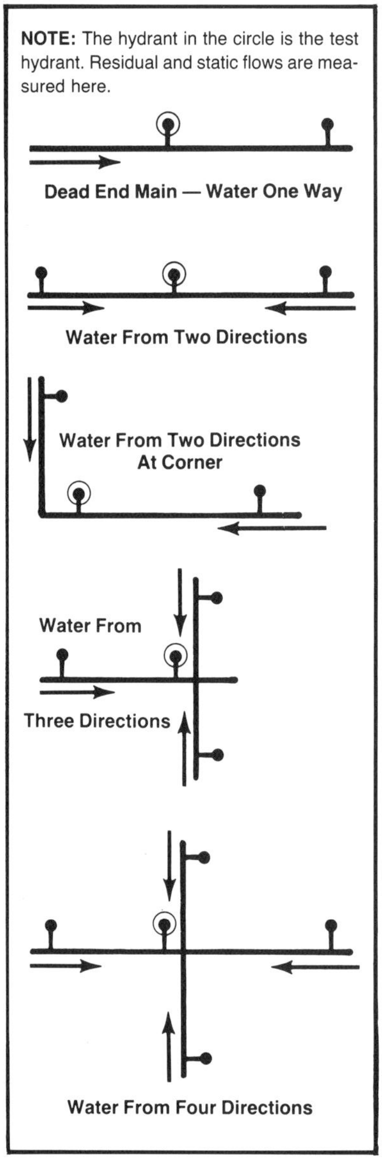

Figure 4.9 There are a number of ways by which fire hydrants in an area can be selected for a flow test.

tion. After checking the second cap for tightness, open the hydrant slowly several turns. Once the air has escaped and a steady stream of water is flowing, the petcock should be closed and the hydrant opened fully.

Step 3: Read and record the static pressure as seen on the pressure gauge.

Step 4: The individual at the flow hydrant(s) should remove the cap(s) from the outlet(s) to be flowed. When using a hydrant outlet, the hydrant coefficient and the actual inside diameter of the orifice should be checked and recorded. If a nozzle is placed on the outlet, its coefficient and diameter should be checked and recorded.

Step 5: Open flow hydrants as necessary, take and record pitot readings of the velocity pressures. The individual at the test hydrant should simultaneously read and record the residual pressure. (**NOTE:** the residual pressure should not drop below 20 psi [138 kPa] during the test. If this happens, the number of flow hydrants must be reduced.)

Step 6: Slowly close the flow hydrants, one at a time, to prevent water hammer in the mains. After checking for proper drainage, replace and secure all hydrant caps. Report any hydrant defects.

Step 7: Check the test hydrant for the return to the normal operating pressure, then close it. The petcock valve should be opened to prevent a vacuum on the gauge. Remove the gauge. After checking for proper drainage, replace and secure the hydrant cap. Report any defects.

FLOW TEST PRECAUTIONS

Certain precautions must be observed before, during, and after conducting flow tests in order to avoid injuries to those participating in the test or to passersby. Efforts must also be made to minimize damage to property from the flowing stream. Both pedestrian and automobile traffic must be controlled during all phases of the testing. This may require assistance from the local law enforcement agency. Other safety measures include tightening caps on hydrant outlets not used, not standing in front of closed caps, and not leaning over the top of the hydrant when operating it. Property damage control measures include opening and closing hydrants slowly to avoid water hammer, not flowing hydrants where adequate drainage is not provided, and remembering to always check downstream to see where the water will flow. Since flowing water across a busy street could cause an accident, take proper measures beforehand to slow or stop traffic. Nor should water be flowed during freezing weather unless street sanding and/or proper drainage minimizes the icing problem. A good rule to follow is: When in doubt — do not flow! If problems exist in making a flow test, thought must be given to their solution so that the test can be completed without disruptions or property destruction.

COMPUTING AVAILABLE FIRE FLOW TEST RESULTS

Graphical

The waterflow chart in Figures 4.10a and b is a logarithmic scale that has been developed to simplify the process of determining available water in an area. The chart is accurate to a reasonable degree if one uses a fine point pencil or pen when plotting results. The figures on the vertical and/or horizontal scales may be multiplied or divided by a constant, as may be necessary to fit any problem.

The procedure for graphical analysis is as follows:

Step 1: Determine which scale should be used.

Step 2: Locate the total water flow measured during the test on the chart.

Step 3: Locate the residual pressure during the test on the chart.

Step 4: Plot the residual pressure above the total water flow measured.

Step 5: Locate and plot the static pressure on the vertical (i.e., at 0 gpm), (0 L/min)

Step 6: Draw a straight line from the static pressure point through the residual pressure point on the waterflow scale.

Step 7: Read the gpm available at 20 psi (138 kPa) and record. This reading represents the total available water that can be relied upon.

Example #1

Test Hydrant (U.S.)

Static 50 psi	Residual 25 psi

Flow Hydrant #1 one 2½" outlet
-pitot reading 7 psi-
C = .9   d = 2.56 in. (2 9/16")
$29.83 \times 0.90 \times (2.56)^2 \times \sqrt{7} = 465.5$ gpm

Flow Hydrant #2 one 2½" outlet
- pitot reading 9 psi-
C = .8   d = 2.44 in. (2 7/16")
$29.83 \times 0.80 \times (2.44)^2 \times \sqrt{9} = \underline{426.2}$ gpm

Total Water Flow = 892 gpm

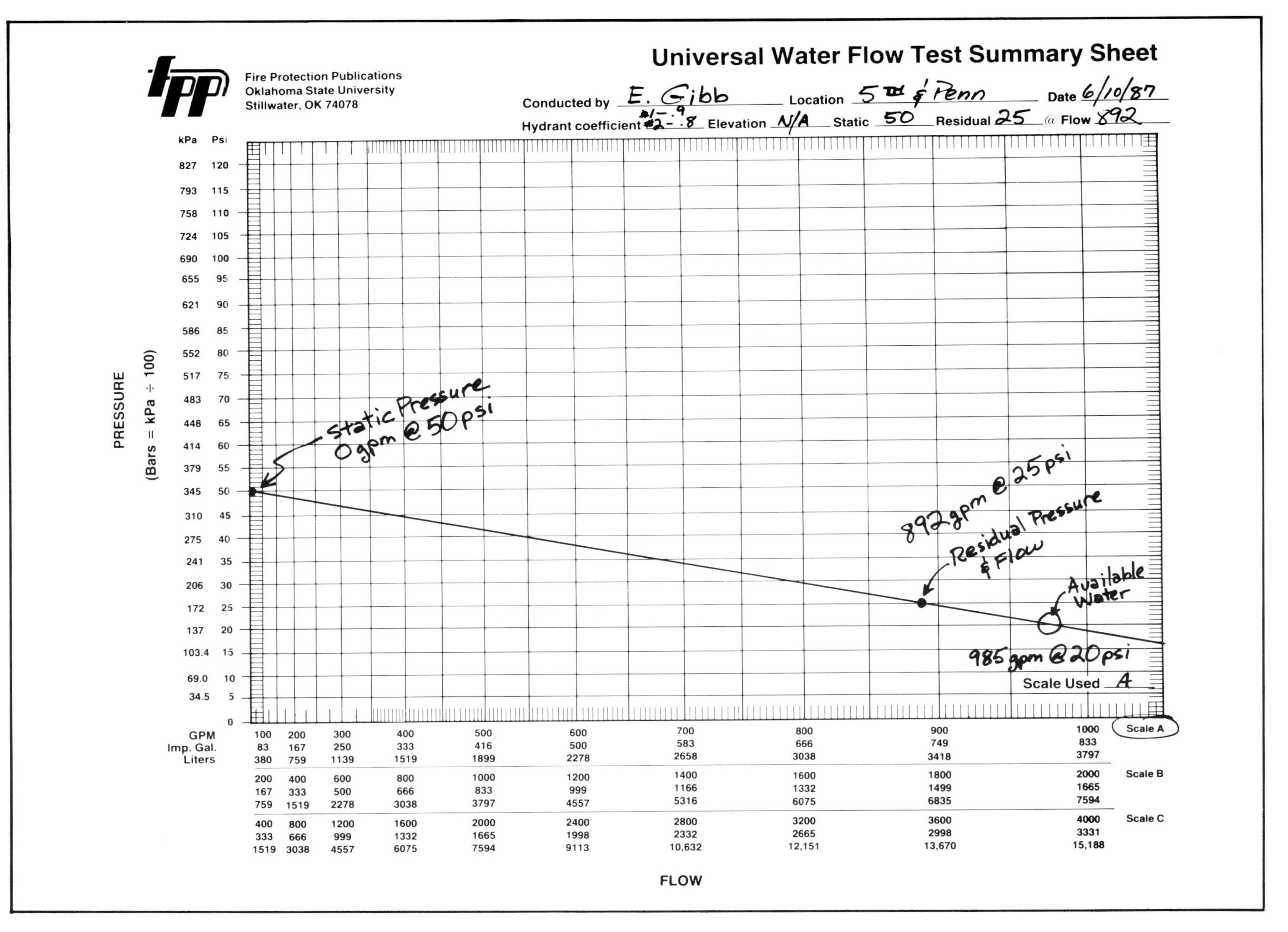

GPM	100	200	300	400	500	600	700	800	900	1000	Scale A
Imp. Gal.	83	167	250	333	416	500	583	666	749	833	
Liters	380	759	1139	1519	1899	2278	2658	3038	3418	3797	
	200	400	600	800	1000	1200	1400	1600	1800	2000	Scale B
	167	333	500	666	833	999	1166	1332	1499	1665	
	759	1519	2278	3038	3797	4557	5316	6075	6835	7594	
	400	800	1200	1600	2000	2400	2800	3200	3600	4000	Scale C
	333	666	999	1332	1665	1998	2332	2665	2998	3331	
	1519	3038	4557	6075	7594	9113	10,632	12,151	13,670	15,188	

Figure 4.10a Flowchart for Example 1 (U.S.).

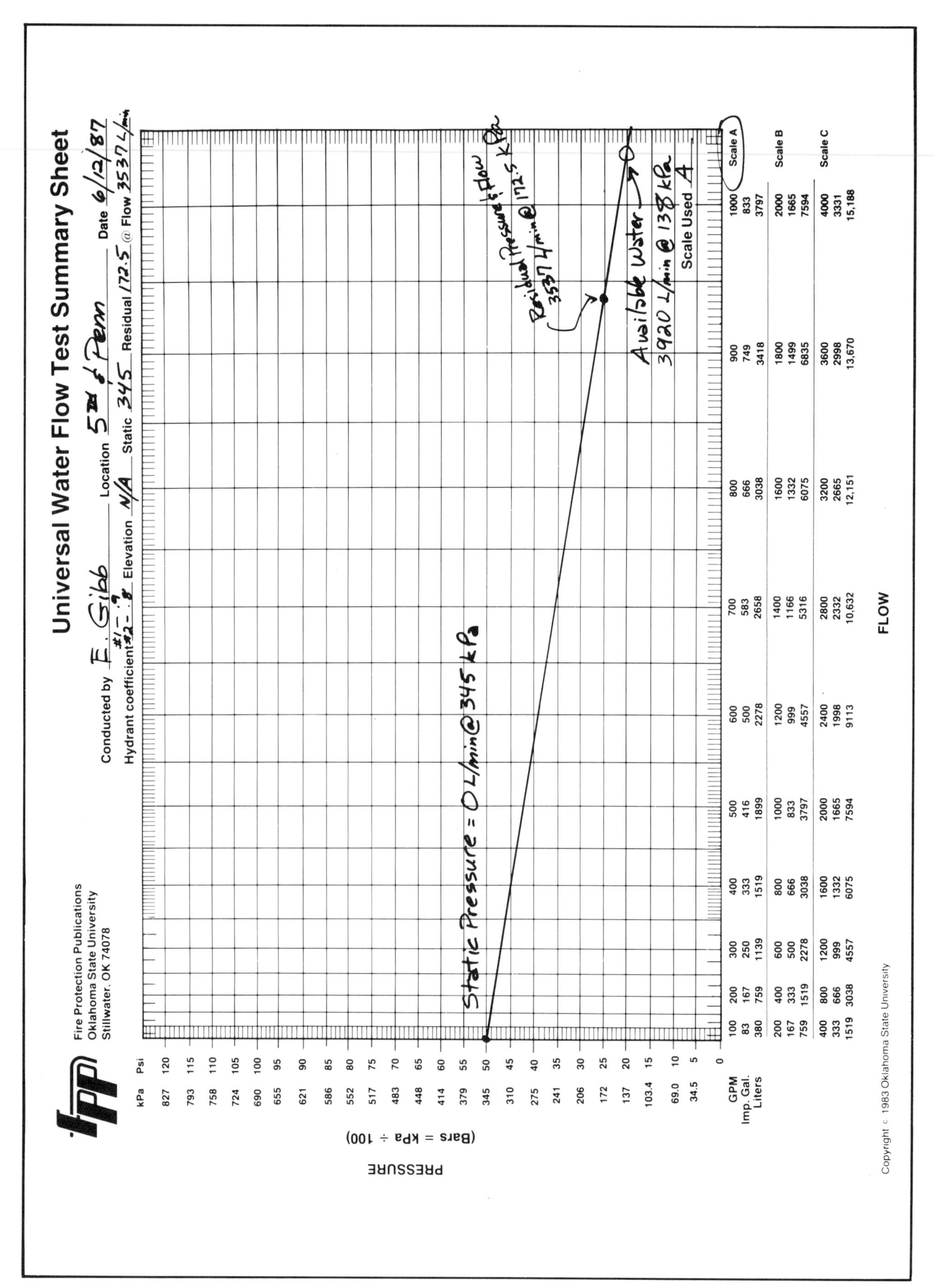

Figure 4.10b Flowchart for Example 1 (metric).

Metric Test Hydrant	Static 345 kPa	Residual 172.5 kPa

Flow Hydrant #1 one 65 mm outlet
- pitot reading 48 kPa-
C = .9   d = 66.5 mm
$0.0667766 \times 0.90 \times (66.5)^2 \times \sqrt{48} = 1\,841.3$ L/min

Hydrant #2 one 2½" outlet
- pitot reading 62 kPa
C = .8   d = 63.5 mm
$0.0667766 \times 0.80 \times (63.5)^2 \times \sqrt{62} = \underline{1\,696.1 \text{ L/min}}$

Total Water Flow = 3 537 L/min

NOTE: The metric formula is not a direct conversion of the U.S. formula.

Figures 4.10a and b show how the test results are plotted for graphical analysis of the water supply. The static pressure of 50 psi (345 kPa) is plotted at 0 gpm (0 L/min). The residual of 25 psi (172.5 kPa) is above the total measured flow of 892 gpm (3 537 L/min) on Scale A. (**NOTE:** It is important to understand that pitot pressures are never plotted on the graph: only the flow that corresponds to the pitot pressures is used). A line drawn through the static and residual pressure points now represents the water supply at the test location. It is easy to note that approximately 985 gpm (3 920 L/min) would be available at 20 psi (138 kPa) or that 675 gpm (2 753 L/min) would be available at 35 psi (241 kPa).

Example #2

Test Hydrant (U.S.)	Static 90 psi	Residual 50 psi

Flow Hydrant 2-2½" flowing
Outlet #1 - 17 psi
Outlet #2 - 17 psi
C = .9   d = 2.56 in.
$29.83 \times .9 \times (2.56)^2 \times \sqrt{17} = 725.4$ gpm
725.4 gpm from each outlet
x 2 outlets = 1,451 gpm

Total Water Flow = 1,451 gpm

Metric Test Hydrant | Static 621 kPa | Residual 345 kPa

Flow hydrant 2-65 mm outlets
Outlet #1 = 117 kPa
Outlet #2 = 117 kPa
C = .9 d = 66.5 mm

$0.0667766 \times 0.9 \times (66.5)^2 \times \sqrt{117} = 2\,874.8$ L/min
2 874.8 L/min from each outlet
2 874.8 x 2 = 5 750

Total Water Flow = 5 750 L/min

NOTE: The metric formula is not a direct conversion of the U.S. formula.

This example shows that the waterflow scale must be changed so a line can be drawn down to the 20 psi (138 kPa) level (Figures 4.11a and b). The available water rate at 20 psi (138 kPa) in this case would be approximately 1,970 gpm (7 825 L/min).

Mathematical Method for Determining Available Water with Hazen-Williams Formula

A variation of the Hazen-Williams formula for determining available water is written:

8\93

$$Q_r = Q_f \times \left(\frac{h_r}{h_f}\right)^{0.54}$$

Q_r = Flow available at desired residual pressure
Q_f = Flow during test
h_r = Pressure drop to residual pressure
h_f = Pressure drop during test

The values of h_r and h_f are supplied in Table 4.3 or can be determined by calculation.

The value of Q_r is supplied from the flow test. Using this information, Q_r can now be found. Using the data from graphical example #1 the following results are obtained:

Q_f = 891.7 gpm
h_r = 50-20, or 30 psi pressure drop, (static pressure minus required residual pressure)
h_f = 50-25, or 25 psi pressure drop, (static pressure minus residual pressure during flow)

By Calculation Method

$$Q_r = 891.7 \left(\frac{30}{25}\right)^{0.54} = 984.0 \text{ gpm}$$

8\93

or

By Table Method

Q_r = 891.7 (1.103) = 984.2 gpm

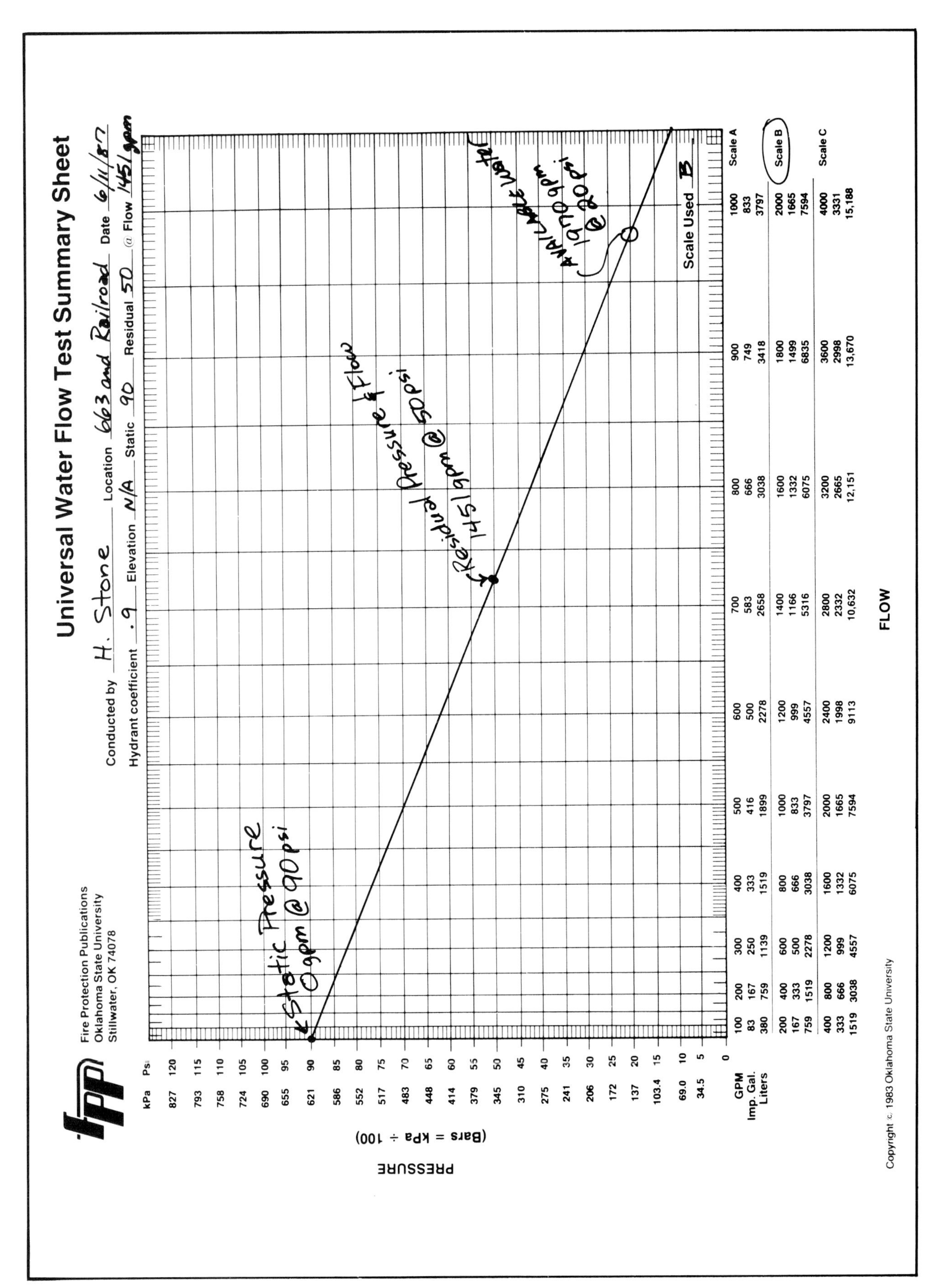

Figure 4.11a Flowchart for Example 2 (U.S.).

5-90

kPa	Psi
827	120
793	115
758	110
724	105
690	100
655	95
621	90
586	85
552	80
517	75
483	70
448	65
414	60
379	55
345	50
310	45
275	40
241	35
206	30
172	25
137	20
103.4	15
69.0	10
34.5	5
	0

GPM	100	200	300	400	500	600	700	800	900	1000	Scale A
Imp. Gal.	83	167	250	333	416	500	583	666	749	833	
Liters	380	759	1139	1519	1899	2278	2658	3038	3418	3797	
	200	400	600	800	1000	1200	1400	1600	1800	2000	Scale B
	167	333	500	666	833	999	1166	1332	1499	1665	
	759	1519	2278	3038	3797	4557	5316	6075	6835	7594	
	400	800	1200	1600	2000	2400	2800	3200	3600	4000	Scale C
	333	666	999	1332	1665	1998	2332	2665	2998	3331	
	1519	3038	4557	6075	7594	9113	10,632	12,151	13,670	15,188	

Figure 4.11b Flowchart for Example 2 (metric).

TABLE 4.3
VALUES OF COMPUTING FIRE FLOW TEST RESULTS

h	$h^{0.54}$	h	$h^{0.54}$	h	$h^{0.54}$	h	$h^{0.54}$	h	$h^{0.54}$	h	$h^{0.54}$	h	$h^{0.54}$
1	1.00	26	5.81	51	8.36	76	10.37	101	12.09	126	13.62	151	15.02
2	1.45	27	5.93	52	8.44	77	10.44	102	12.15	127	13.68	152	15.07
3	1.81	28	6.05	53	8.53	78	10.51	103	12.22	128	13.74	153	15.13
4	2.11	29	6.16	54	8.62	79	10.59	104	12.28	129	13.80	154	15.18
5	2.39	30	6.28	55	8.71	80	10.66	105	12.34	130	13.85	155	15.23
6	2.63	31	6.39	56	8.79	81	10.73	106	12.41	131	13.91	156	15.29
7	2.86	32	6.50	57	8.88	82	10.80	107	12.47	132	13.97	157	15.34
8	3.07	33	6.61	58	8.96	83	10.87	108	12.53	133	14.02	158	15.39
9	3.28	34	6.71	59	9.04	84	10.94	109	12.60	134	14.08	159	15.44
10	3.47	35	6.82	60	9.12	85	11.01	110	12.66	135	14.14	160	15.50
11	3.65	36	6.93	61	9.21	86	11.08	111	12.72	136	14.19	161	15.55
12	3.83	37	7.03	62	9.29	87	11.15	112	12.78	137	14.25	162	15.60
13	4.00	38	7.13	63	9.37	88	11.22	113	12.84	138	14.31	163	15.65
14	4.16	39	7.23	64	9.45	89	11.29	114	12.90	139	14.36	164	15.70
15	4.32	40	7.33	65	9.53	90	11.36	115	12.96	140	14.42	165	15.76
16	4.47	41	7.43	66	9.61	91	11.43	116	13.03	141	14.47	166	15.81
17	4.62	42	7.53	67	9.69	92	11.49	117	13.09	142	14.53	167	15.86
18	4.76	43	7.62	68	9.76	93	11.56	118	13.15	143	14.58	168	15.91
19	4.90	44	7.72	69	9.84	94	11.63	119	13.21	144	14.64	169	15.96
20	5.04	45	7.81	70	9.92	95	11.69	120	13.27	145	14.69	170	16.01
21	5.18	46	7.91	71	9.99	96	11.76	121	13.33	146	14.75	171	16.06
22	5.31	47	8.00	72	10.07	97	11.83	122	13.39	147	14.80	172	16.11
23	5.44	48	8.09	73	10.14	98	11.89	123	13.44	148	14.86	173	16.16
24	5.56	49	8.18	74	10.22	99	11.96	124	13.50	149	14.91	174	16.21
25	5.69	50	8.27	75	10.29	100	12.02	125	13.56	150	14.97	175	16.26

Metric

Q_r = 3537.4 L/min

h_r = 345 kPa - 137 kPa, or 208 kPa pressure drop (static pressure minus required residual pressure)

h_f = 345 kPa - 172.5 kPa, or 172.5 kPa pressure drop (static pressure minus residual pressure during flow)

$$Q_r = \frac{(208)^{0.54}}{(172.5)^{0.54}} = 3913.6 \text{ L/min}$$

Determination of Available Water: Nomograph

The flow available at any desired residual pressure can also be determined by nomograph (Figures 4.12a and b) This nomograph is based on the Hazen-Williams derivation used in the mathematical method of determining available water.

The procedure for nomograph analysis is the following:

Determine h_r = allowable pressure drop

Determine h_f = observed pressure drop

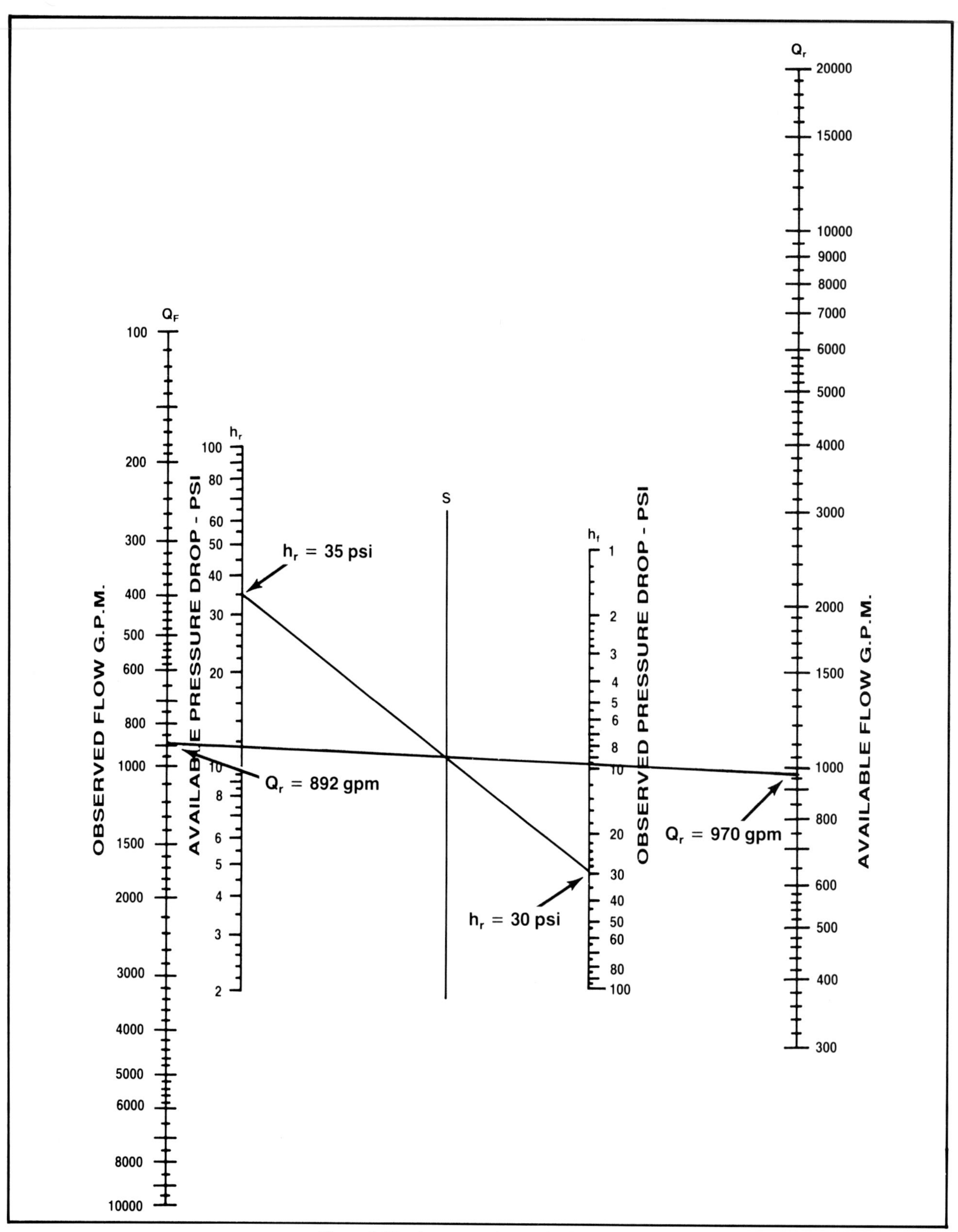

Figure 4.12a A nomograph can also be used to determine water flow available at a particular pressure. The example shown uses U.S. measurements.

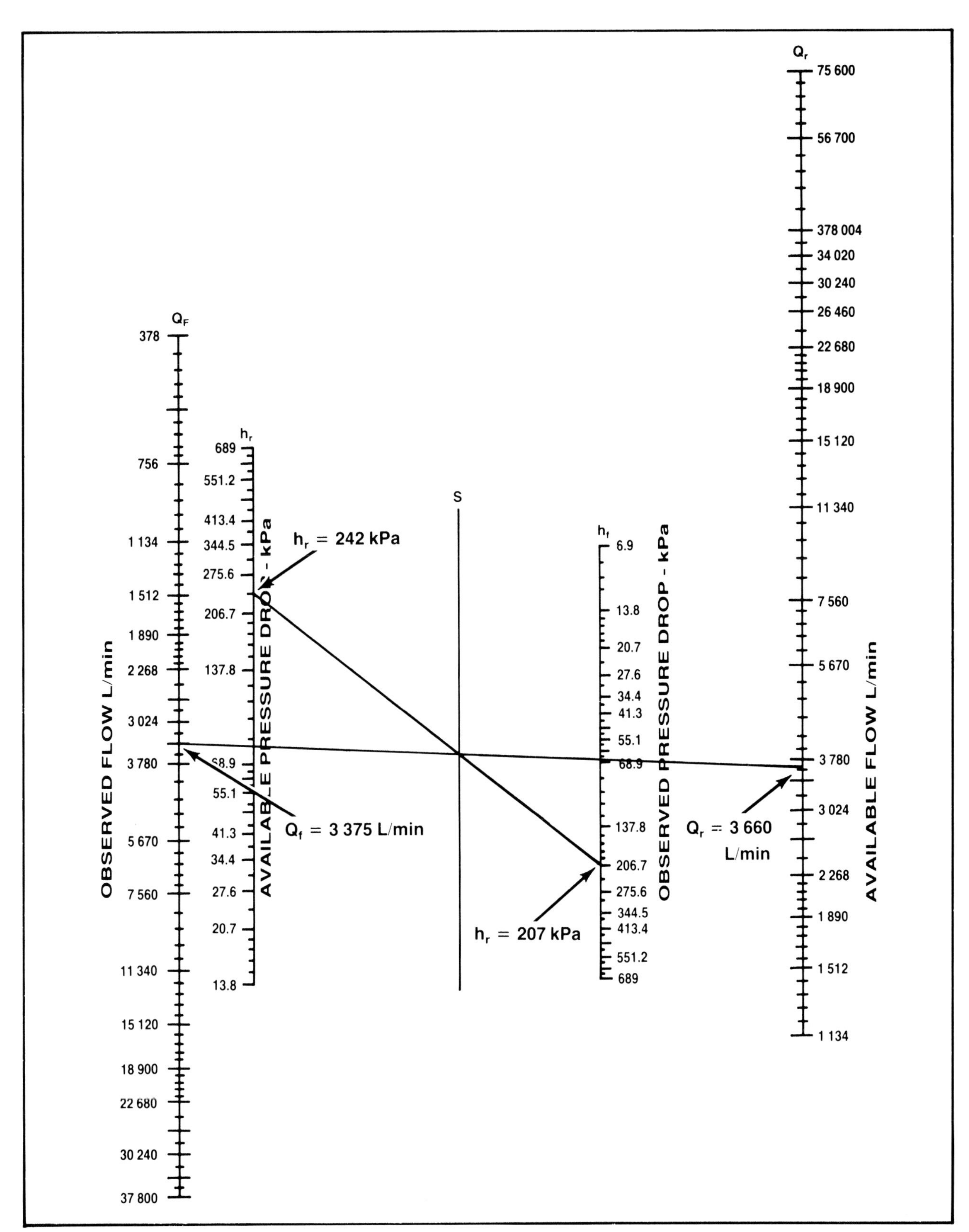

Figure 4.12b An example of a nomograph using metric measurements.

Step 1: Using a straightedge, connect points h_r and h_f. Using a sharp point pencil or pen, mark the intersection on line S.

Step 2: Rotate the straightedge about the intersection point on S until it is in line with the observed flow on Q_f scale.

Step 3: Where this line extends, mark the intersection on the Q_r scale.
This Q_r point is the available flow at the assumed pressure drop.

Once again using the data from graphical example #1, the following results are obtained:

Q_f = 891.7 gpm
h_r 50-20, or 30 psi (allowable pressure drop)
h_f 50-25, or 25 psi (observed pressure drop)

Lining up h_r and h_f, the intersection point on line S is obtained. Rotating the straightedge about this point till it lines up with the observed flow (891.7 gpm [3 375.4 L/min]), the available flow is determined by extending the line to scale (Q_r) ($\approx$ 980 gpm [3 710 L/min]).

It is now clear that several methods are available to determine the available water flow resulting from the test as well as the available flow at any system pressure. With this information, several questions can be addressed:

- Does the water system meet the needed fire flow for the occupancy or area?
- Is the available system flow adequate to provide sprinkler system demands in the area?
- Where will water system improvements be needed?

It can be noted that any of the three available waterflow determination methods gives flows that are almost numerically identical.

Each type of determination method has its own advantages. Graphical analysis provides a visual graph that indicates system capability at any pressure. The nomograph is often useful for field measurements and provides close approximations. Mathematical determination provides the most exact answer for any given situation and may be desirable for use in computer applications where large amounts of flow test data must be analyzed. Fire departments should choose their analysis method based on individual needs. Discrepancies found in some of the metric examples are due to the rounded off factors used currently as well as those in use during the printing of the Universal Water Flow Test Summary Sheet.

Chapter 4 Review

Answers on page 261

TRUE-FALSE. Mark each statement true or false. If false, explain why.

1. The principal reason for conducting fire flow tests is to flush the water piping system to ensure maximum flow when needed.

   ☐ T ☐ F ____

   ____

2. A pitot tube is used to record the static pressure found in a water system.

   ☐ T ☐ F ____

   ____

3. When flowing only one hydrant, the test hydrant should be in between the water source and the flow hydrant.

   ☐ T ☐ F ____

   ____

4. A small column of water remains in the barrel of a dry-barrel hydrant after it is closed.

   ☐ T ☐ F ____

   ____

MULTIPLE CHOICE: Circle the correct answer.

5. Identify the first step in conducting a fire flow test.
   A. Examine hydrant for physical damage
   B. Inspect threads for burrs or other damage
   C. Remove overgrown brush from around the hydrant
   D. Notify the water department

6. When flow testing hydrants, the minimum amount the static pressure should be made to drop is __________.
   A. 5 psi (35 kPa)
   B. 10 psi (69 kPa)
   C. 15 psi (105 kPa)
   D. 20 psi (138 kPa)

7. What is the minimum residual pressure required to compute available water for an area?
   A. 5 psi (35 kPa)
   B. 10 psi (69 kPa)
   C. 15 psi (105 kPa)
   D. 20 psi (138 kPa)

8. Why will the $29.83 \times C \times d^2 \times \sqrt{P}$ formula alone not work when using a 4½-inch (115 mm) outlet?
   A. Voids in the water stream
   B. The formula is designed for 2½-inch (65 mm) outlets only
   C. The flow is beyond the ability of the formula
   D. There are no coefficients for large diameter outlets

9. Identify the *two* problems below that may be a result of pressure differentials in a water main.
   A. Main collapse
   B. Discoloration of the water
   C. Damage to testing equipment
   D. Pump cavitation

10. What is the easiest method of determining the amount of water being discharged from a hydrant butt?
    A. Hazen-Williams Formula
    B. Nomograph
    C. Prepared table or chart
    D. A computer program

SHORT ANSWER: Answer each item briefly.

11. State the proper formula for determining the flow from a hydrant butt and tell what each variable stands for.

____

____

____

____

____

12. During a hydrant flow test, 85 psi (586 kPa) is measured from a 2½-inch (65 mm) outlet with a coefficient of discharge of .90. What is the flow in gpm (L/min)?

____

____

13. What are at least two questions that can be addressed after finishing a complete waterflow analysis?

____

____

14. Explain what the coefficient of discharge is and why it is important.

____

____

Photo by Ronald Jeffers

Chapter 5

Static Sources

This chapter provides information that addresses performance objectives described in NFPA 1002, *Fire Apparatus Driver/Operator Professional Qualifications* (1988), particularly those referenced in the following section:

3-2.11

Chapter 5
Static Sources

Most urban areas are covered by a network of hydrants and water mains that make up the public water distribution system. Rural areas often do not have this convenience and must depend on other means and sources to provide sufficient water for fire fighting. Even in areas served by a network of mains, the domestic supply may be limited and mains far too small to provide adequate fire flows.

Portable and static water supplies are considered to be the primary resource for providing rural fire protection water. Urban fire departments must also consider these sources in the event of water system failure, as an auxiliary supply to augment the water system at large fires or in weak areas. Emergency operation plans should be developed in case of a natural disaster or man-made interruption of the water system. Once auxiliary sources are identified and contingency plans prepared, the fire department should train in utilizing these sources so problem areas can be corrected before any water emergency occurs.

NATURAL WATER SOURCES

Rivers and streams are a common source of supplementary water (Figure 5.1). They provide a flowing source for drafting op-

Figure 5.1 Rivers generally make an excellent static source of water.

erations. The flow capacity of a river or stream must be evaluated to determine its merit as a fire protection water source (Figure 5.2). A stream 10 feet (3 m) wide with an average depth of 1 foot (0.3 m) flowing approximately 15 feet per minute (4.6 m/min) will supply 1,122 gpm (4 247 L/min). This is determined by the formula:

$$Q = A \times V$$
$$Q = \text{Flow in ft}^3\text{/min}$$
$$A = \text{Area in ft}^2$$
$$V = \text{Velocity in ft/min}$$
$$\begin{aligned} Q &= A \times V \\ &= 10 \text{ ft} \times 1 \text{ ft} \times (15 \text{ ft/min}) \\ &= 150 \text{ ft}^3\text{/min} \end{aligned}$$
$$150 \text{ ft}^3\text{/min} \times 7.48 \text{ gal/ft}^3 = 1{,}122 \text{ gpm}$$

Metric:

$$Q = A \times V$$
$$Q = \text{Flow in m}^3\text{/min}$$
$$A = \text{Area in m}^2$$
$$V = \text{Velocity in m/min}$$
$$\begin{aligned} Q &= A \times V \\ &= (3 \text{ m} \times 0.3 \text{ m}) \times (4.6 \text{ m/min}) \\ &= 4.14 \text{ m}^3\text{/min} \\ &= 4.14 \text{ m}^3\text{/min} \times 1\,000 \text{ L/m}^3 = 4\,140 \text{ L/min} \end{aligned}$$

This formula is also used for determining water flow in piping. However, it is not practical for fire hose applications since velocity of the water stream is difficult to determine.

Freeman's formula used in Chapter 4 is a derivation of Q = A x V.

Figure 5.2 Provided they are accessible and capable of flowing reasonable amounts of water, small streams may be used as a water supply source.

Lakes and ponds also provide water for fire protection (Figure 5.3). As with other sources, the capacity of lakes and ponds is crucial to their use as a fire fighting water source. A rule of thumb for evaluating lake capacity is that every 1 foot (0.3 m) of depth for an area of 1 acre (0.4 ha) (approximately the size of a football field) will provide 1,000 gpm (3 785 L/min) for 5 hours (Figure 5.4).

Figure 5.3 Many rural homeowners build ponds on their properties to provide water for livestock and fire protection.

Figure 5.4 Lakes provide virtually endless static supply of water. *Courtesy of Rich Mahaney.*

Accessibility/Reliability

The major problems that occur with natural sources are poor accessibility and unreliability. The conditions that cause these problems are as follows:

- Inability to reach water with pumper (Figure 5.5)

Figure 5.5 A static supply of water may not be usable due to its location. The scale attached to this bridge shows that the water is too far below the bridge to be drafted from by fire apparatus.

- Wet or soft ground approaches
- Inadequate depth for draft
- Silt and debris
- Freezing weather (Figure 5.6)
- Drying up (Figure 5.7)

Figure 5.6 In colder climates, firefighters may have to break through ice in order to use a static water supply source. *Courtesy of Rich Mahaney.*

Figure 5.7 During dry spells, static water sources may not be at normal levels. The water shown above would be inaccessible from the right side due to the unstable, exposed river bed.

Areas will often be encountered where pumpers will not be able to reach the water due to the type of terrain encountered. Weather conditions can also limit access to static sources. Fire departments can avoid this problem by utilizing public boat launching facilities by constructing gravel drives, or clearing brush to drafting points of high usage (Figure 5.8). Bridges may provide a useful drafting location, but must be checked for adequate width, height, and load requirements for fire apparatus and personnel safety. Another consideration is that pumpers can generally only

Figure 5.8 Many recreational waterways are equipped with boat launching facilities that make excellent access points for drafting operations.

pull a 15 to 20 foot (4.6 m to 6 m) draft to the centerline of the pump. This will also reduce the pump capacity to 70 percent at 15 feet (4.6 m) and 60 percent at 20 feet (6 m).

Wet ground approaches are a hazardous area for fire department apparatus. Grass and vegetation can hide soft ground spots, or after a vehicle stops for a period of time, settling may occur. Wet approaches can trap apparatus, effectively blocking access to the water source. Personnel and other apparatus will have to be called in to free the stuck vehicle.

The depth for drafting available at a water source is an important operational consideration. A depth of 24 inches (610 mm) of water above and below the hard suction strainer is a good rule of thumb for minimum draft depth (Figure 5.9), although lesser

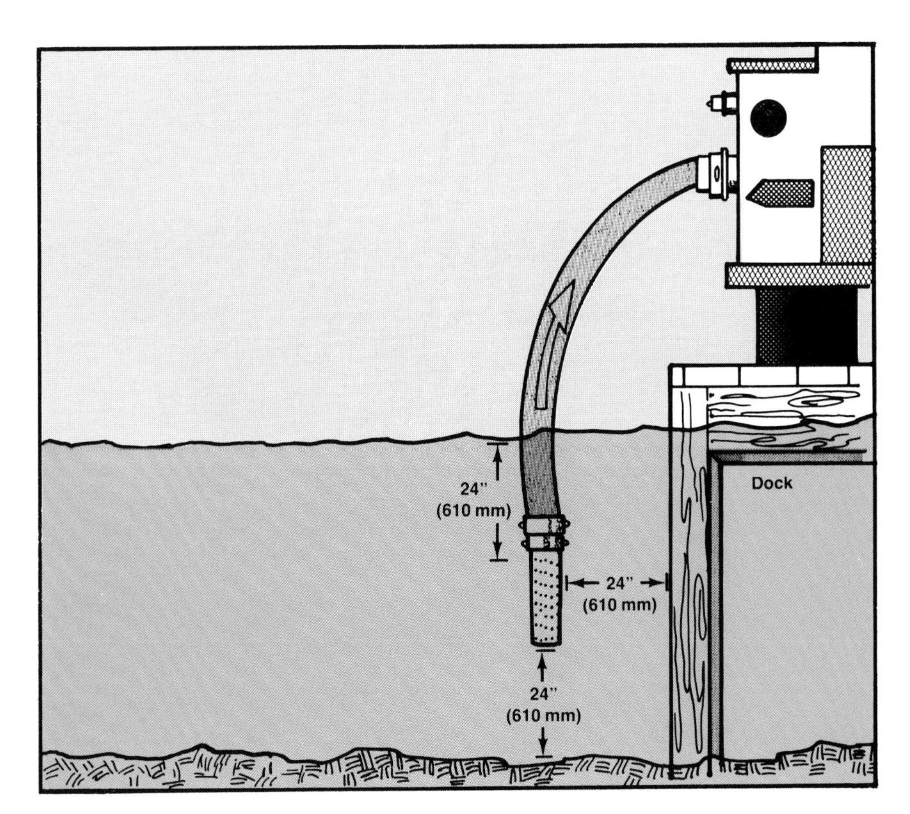

Figure 5.9 A minimum of 24 inches (610 mm) of water should surround all sides of a strainer.

depths have been used successfully. Special drafting or floating strainers are available for use in shallow sources and can draw water down to a 1- to 2-inch (25 mm to 51 mm) level (Figures 5.10a and b). At locations where the minimum draft depth is not adequate, several modifications are available to create and maintain sufficient depth:

- Diking and damming of streams (Figure 5.11).
- Dry hydrant (installed below pond or stream bed).
- Draft access basins (basin or cistern adjacent to rivers, ponds, and streams).

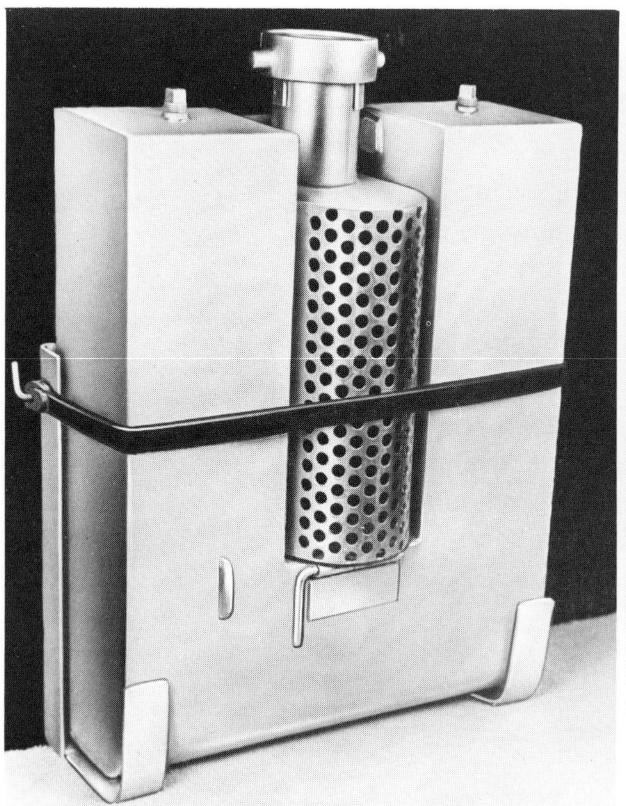

Figure 5.10b A floating dock strainer in operation. *Courtesy of Rich Mahaney.*

Figure 5.10a A bottom view of a floating dock strainer.

Figure 5.11 By damming this stream, firefighters have created a drafting point that, with the aid of a floating pump, is capable of supplying an effective fire stream.

When considering any modification of sources, the local governing agency or property owner should be consulted before any work is performed. A structural analysis of major changes may also be required to determine the strength and safety of the change.

Silt and debris can render a source useless by clogging strainers, by seizing-up or damaging pumps, and by allowing sand and small rocks to enter attack lines and clog fog stream nozzles. All hard suction lines should have strainers on them whenever drafting from a natural source. The suction hose should be located and supported so the strainer does not rest on or near the bottom. A preferable method for avoiding silt and debris is by the installation of a dry hydrant. A dry hydrant allows easy access to natural sources without the set-up time required for a regular drafting operation. They also avoid the wet ground approach problem and can circumvent the problem of silt and debris when installed properly. Dry hydrants employ 6 inch (150 mm) or larger pipe run from a local large hose threaded connection to a below water level intake and strainer with a gravel bed to block out silt (Figure 5.12).

Courtesy of Ken-Mar, Inc.

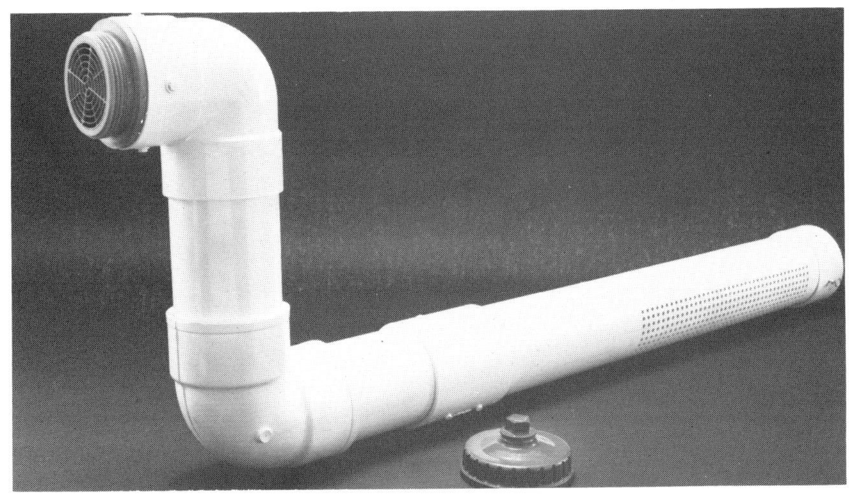

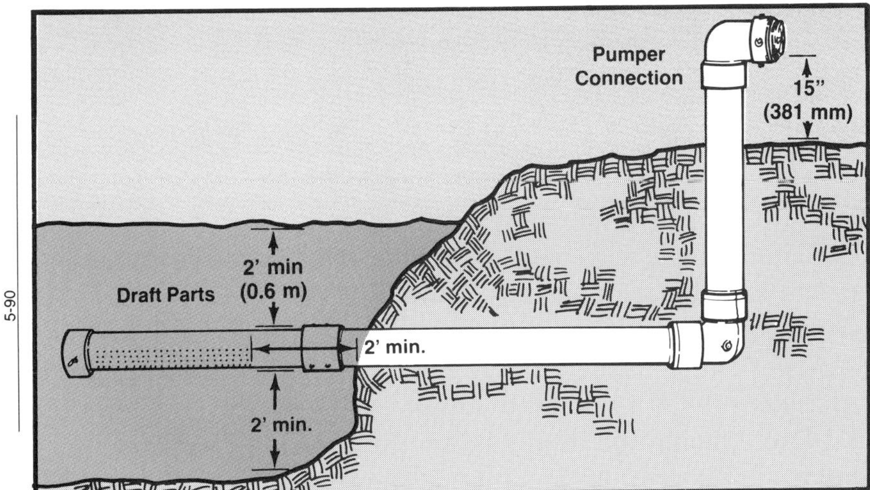

Figure 5.12 Dry hydrants may be installed at locations frequently used for drafting operations.

Regardless of intake method, any operation that causes salt water or dirty water to be drawn into the pump requires that the pump be flushed with clean fresh water after the water supply operation is completed. This will help prevent corrosion to the pump and its components.

Figure 5.13 In areas subject to freezing, a floating barrel may be positioned in the water at a drafting site. This will speed drafting operations when the water is frozen.

A major problem in many parts of the country is that winter freezing can effectively block the majority of static supplies. Surface freezing can make access difficult at best, while shallow ponds and lakes may freeze to near bottom, rendering them useless. Methods that can be used to aid access to surface frozen ponds and lakes are barrels with antifreeze solution floated on the water surface or wooden plugs or plastic garbage cans stabilized at a location so they may be driven through the ice once the source has frozen (Figure 5.13). Both these operations require the fire department to have these devices in place before the seasonal freeze. Sources that do not have thick ice over them can be broken open with an axe, pike pole, or drilled through with hand augers. Firefighters must take care not to weaken the ice around these openings and should be prepared for an ice rescue situation. Power augers can be used to drill the thicker ice coverings (Figure 5.14).

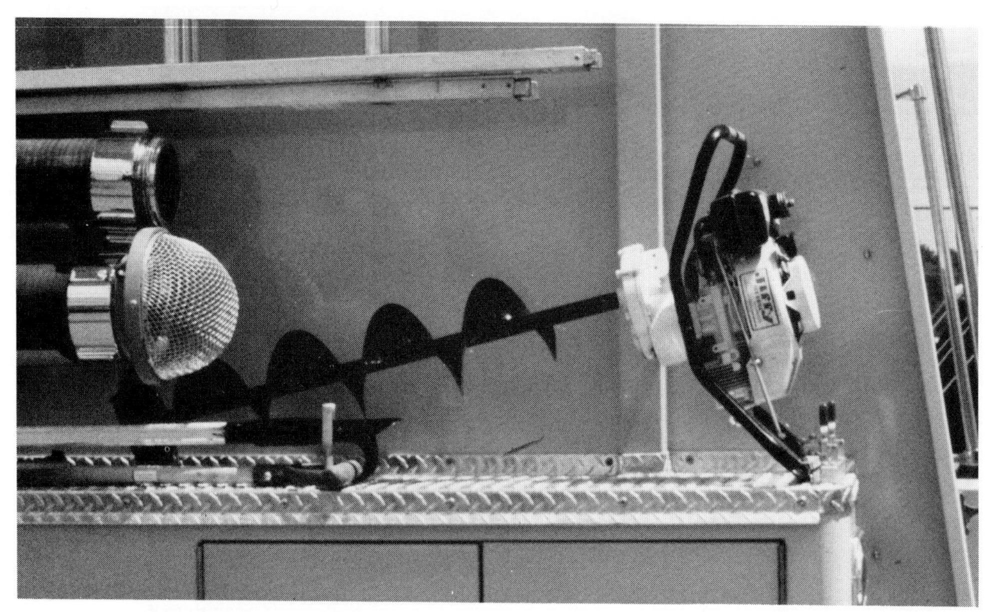

Figure 5.14 Power augers may be carried on apparatus to open drafting holes in the ice. *Courtesy of Rich Mahaney.*

MAN-MADE SOURCES

Cisterns have been used to provide extra water for domestic use and fire fighting for many years (Figure 5.15). They should have adequate water capacity for the area they serve as well as the proper connections for the pumper. Freezing can be a problem with cisterns as it is with lakes and ponds if they are not below the frost line. Placing the intake below the frost level, in a manner similar to the dry hydrant intake placement, will allow year-round access to the source.

Domestic water tanks and farm tanks may hold plenty of water for fire protection but the fire department liaison must meet with owners to see that access and connections are available for fire department use (Figure 5.16).

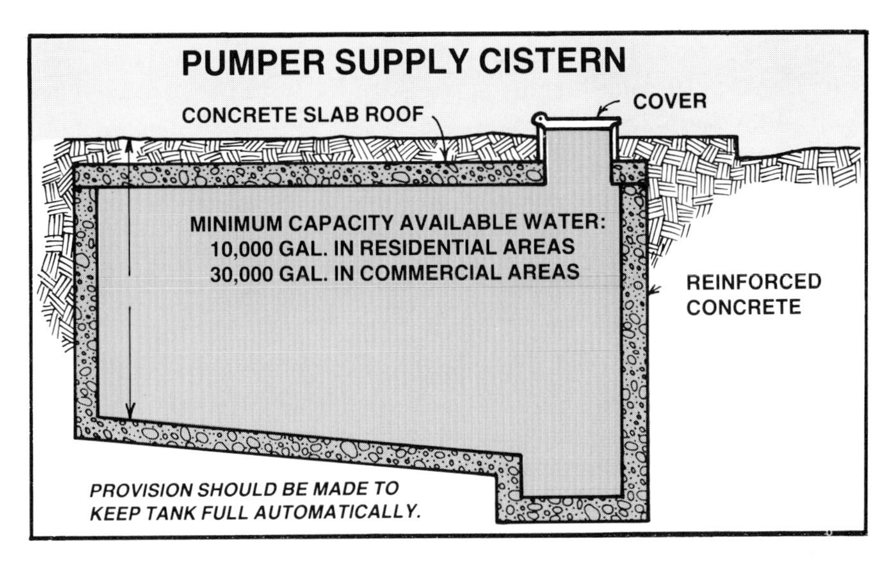

Figure 5.15 Shown are recommended specifications for pumper supply cisterns.

Figure 5.16 Some private property owners may have their own small water tank, which may or may not be elevated.

Swimming pools provide an excellent source of clean water for fire fighting. Pools have a unique access problem since they are often fenced for security purposes (Figure 5.17). Fire departments that have a high concentration of pools in their protection district should make arrangements to gain access to the pool. Apparatus operators must also take extra care when trying to reach the access points of swimming pools. Individual plans may be required for pools with poor access or questionable support around their edges. Floating or stationary portable pumps may often be used for supply when pumpers are unable to reach the pool (or other water source). The capacity of a pool can be calculated by one of the following formulas:

Square Pool

Capacity in gallons = L x W x D x 7.5

Where: L = Length in feet
W = Width in feet
D = Average depth in feet
7.5 = Number of gallons per cubic foot

Capacity in liters = L x W x D x 1000

Where: L = Length in meters
W = Width in meters
D = Average depth in meters
1,000 = Number of liters per cubic meter

Figure 5.17 Swimming pools are often inaccessible to apparatus and require the use of portable pumps to feed water to pumpers.

Round Pool

Capacity in gallons = $\pi \times r^2 \times D \times 7.5$

Where: Pi(π) = 3.14
r = radius or ½ the diameter in feet
D = Average depth in feet
7.5 = Number of gallons per cubic foot

Capacity in liters = $\pi \times r^2 \times D \times 1000$

Where: Pi(π) = 3.14
r = radius or ½ diameter in meters
D = Average depth in meters
1,000 = Number of liters per cubic meter

Two other considerations when using swimming pools are: that they may be drained during the winter months, and drafting next to the fire building may pose an exposure hazard to equipment and personnel.

Agriculture irrigation systems are another potential fire protection water source (Figure 5.18). Systems in some locations may flow well in excess of 1,000 gpm (3 785 L/min). Irrigation water is generally transported in two methods: open canals and piped systems. Open canals may run through a property and may have several accessible points that pumpers can draft from. Depth of canals may be such that special floating strainers must be used. Piped systems are the more common type of irrigation water system. As with canals, several points may be accessible to fire department pumpers, but the piping must have connections that are usable at these locations. Special threaded adapters and specialized tools may be needed to utilize an irrigation system.

Figure 5.18 This irrigation pump has a fire department connection for emergency use.

Man-made water sources can have the same deficiencies as natural sources. Access for drafting may be as difficult as with natural sources. These water sources may be scattered throughout a property; therefore, locating them and determining their accessibility must be accomplished before a fire occurs.

Cisterns, wells, tanks, and irrigation systems can all be subject to freezing. The fire department must determine the systems that are designed for year-round use and the ones that will have to be maintained by the fire department.

Pre-incident planning is important to ensure availability and reliability of static water sources. Searching for water sources while responding to an alarm will delay setup of water supply operations and may jeopardize the fire suppression effort (Figure 5.19).

Areas may exist where water sources are simply unavailable. In these instances, the fire department's tanker and other types of tank vehicles become the water sources (Figure 5.20 on page 132). Fire departments must be ready to call on any and all water haulers when drought conditions or a situation of minimum water threatens.

A method for determining the water supply requirements in areas not covered by a water distribution system is outlined in NFPA 1231, *Water Supplies for Suburban and Rural Firefighting*.

NFPA 1231 provides a consensus standard for determining the total water supply and fireground delivery rate for rural incidents. The standard differs from the Insurance Services Office Fire Flow Guide in the following manner:

- ISO — Determines necessary flow rate from the water system required to protect and control fires for a given area, or specific occupancy.
- NFPA 1231 — Determines the total quantity of water in gallons (Liters) that will be required for a given occupancy and the corresponding fireground application rate necessary for fire control.

Excerpts of the segments directly related to determining the water supply for a given occupancy are reprinted as Appendices B and C. The entire standard, NFPA 1231, can be referenced for additional information.

Special considerations about the use of NFPA 1231 are that the standard is not intended to provide details for calculating an adequate amount of water for large special fire protection problems such as bulk petroleum storage, bulk flammable gas storage, large varnish and paint factories, and so on. When the water supply is being calculated for severe hazard occupancies, the fire de-

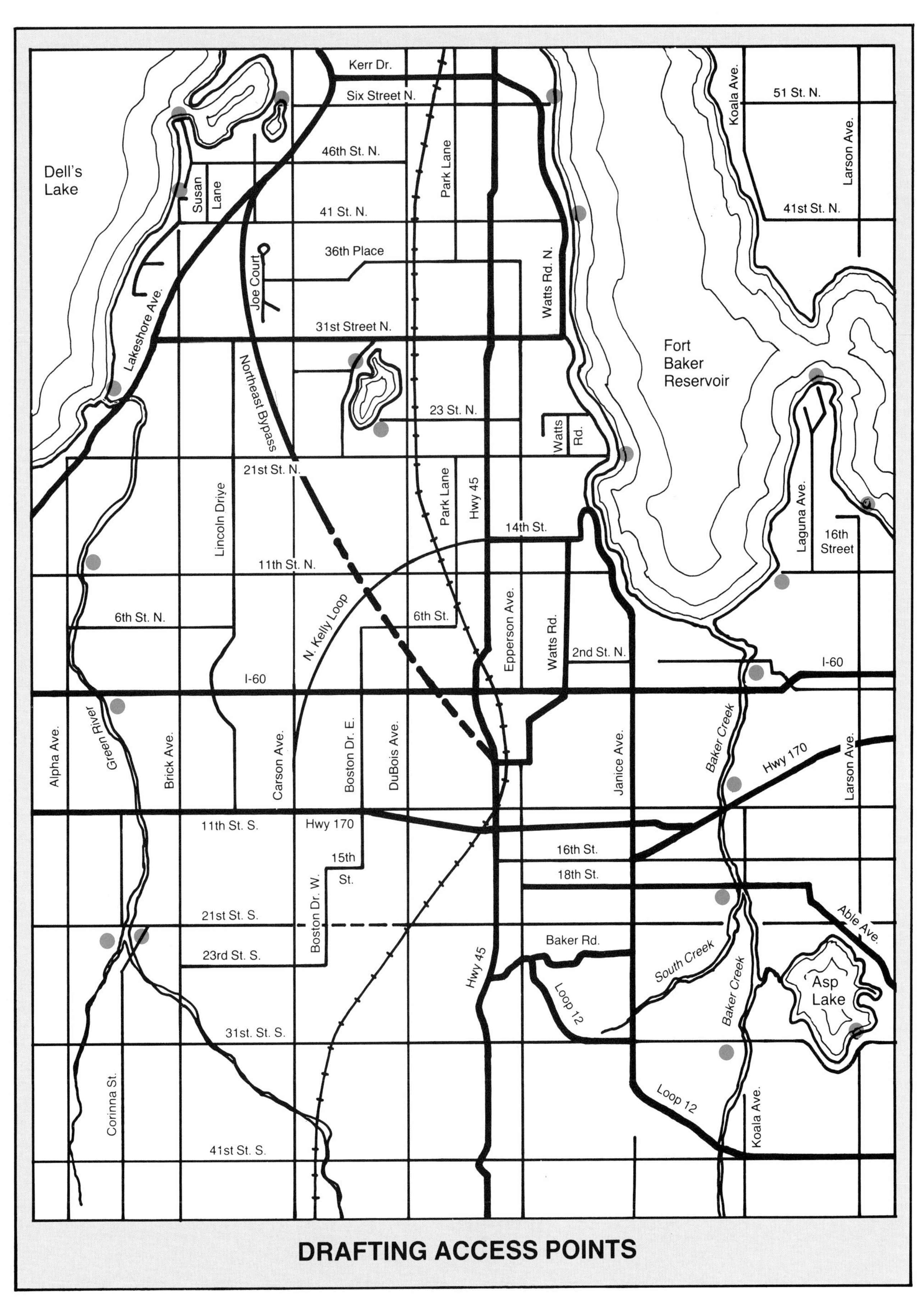

Figure 5.19 A map of static water supplies is a great aid to responding fire units.

Figure 5.20 Fire department tankers provide their own water supply. *Courtesy of Joel Woods.*

partment may need to revise water supply and delivery rates when the hazard is determined excessive. Separate NFPA standards may apply to these occupancies and the fire department should consult the respective standard for additional water supply requirements.

The fire department may waive the requirements of this standard for any occupancy totally protected by an automatic sprinkler system installed in compliance with NFPA 13, *Standard for the Installation of Sprinkler Systems.* Systems installed under NFPA 13 have already met water supply requirements for additional exterior hoselines.

Chapter 5 Review

Answers on page 261

TRUE-FALSE. Mark each statement true or false. If false, explain why.

1. To obtain the best results when using a hard suction strainer, allow the strainer to rest on the bottom.

   ☐ T ☐ F ____

   ____

2. Regardless of the flow rate, a small stream with less than 1 foot (0.3 m) of water will never be sufficient to support drafting operations.

   ☐ T ☐ F ____

   ____

MULTIPLE CHOICE: Circle the correct answer.

3. Which of the following would *not* be considered a static water supply source?
   A. Elevated storage tank
   B. Pond
   C. Lake
   D. Cistern

4. Identify the major problem associated with using swimming pools as a water supply source.
   A. Chemicals
   B. Potential damage to the pool liner
   C. Small children playing in the area
   D. Poor accessibility

5. What is the maximum height that a pump can generally pull draft?
   A. 10 feet (3 m)
   B. 12 feet (4 m)
   C. 20 feet (6 m)
   D. 23 feet (7 m)

FILL IN THE BLANK: Fill in the blanks with the correct response.

6. When drafting utilizing a hard suction strainer, __________ inches/mm of water should be above the strainer.

7. The two transportation methods for water in irrigation systems are open canals and __________.

SHORT ANSWER: Answer each item briefly.

8. A 15-foot (4.57 m) wide stream is moving at 8 feet (2.43 m) per minute. The stream is 2 feet (0.61 m) deep. What is the flow capacity of the stream?

   ____

   ____

9. List at least four conditions that may affect the accessibility or reliability of natural sources of water.

10. A 22-foot (6.7 m) diameter swimming pool has an average depth of 4½ feet (1.37 m). How much water is available for use from this pool?

11. Describe the rule of thumb for evaluating the water capacity of a lake.

12. Describe several methods by which bodies of water can be prepared to aid access during freezing conditions.

13. How can the problem of recurrent wet access to water supplies be dealt with?

Chapter 6

Photo by Bill Eckman

Relay Operations

Chapter 6
Relay Operations

In many fire situations, it is impossible to locate the attack pumper at the water source. In order to supply water to the fire scene, relay operations must be employed from the water source.

A relay utilizes a pumper at the source location to supply water under pressure through a number of hoseline combinations to the next in-line pumper. This pumper, in turn, boosts the pressure to supply the next pumper, and so on, until water reaches the fireground apparatus. The only limiting factors to a relay are the number of apparatus available to overcome friction loss in the supply lines, the amount of hose available to complete a relay, and a reliable source.

APPARATUS

Apparatus used in water relays consist of pumpers that are assigned to various functions within a relay (Figure 6.1 on next page).

- The source pumper will be connected to a hydrant supply or will draft from a static supply. In either case, the pumper will then supply water through supply lines to the next in-line pumper.
- The in-line pumper is a pumper or pumpers connected within the relay that receives water from the source pumper or another in-line pumper, boosts the pressure, and then supplies water to the next in-line pumper or the attack pumper.
- The attack pumper is a pumper located at the fire scene that will be receiving water from the relay and supplying attack lines and appliances as needed for fire suppression or as governed by the available water supply.

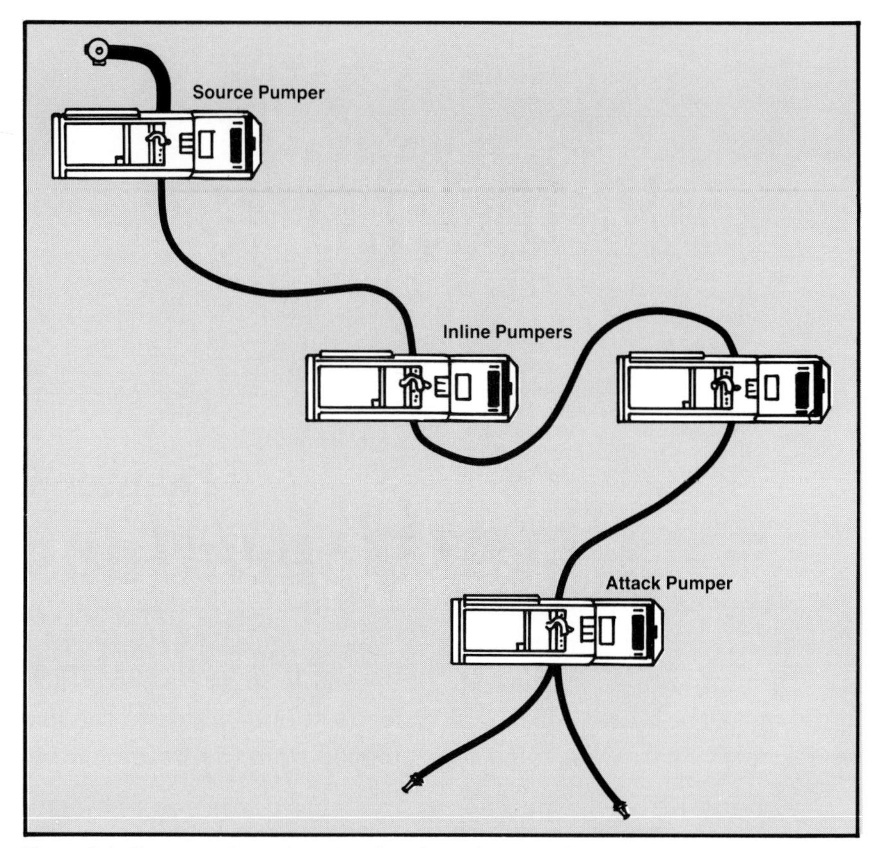

Figure 6.1 Shown are the various functions in a relay operation.

Another form of apparatus specifically used in relay operations is the relay reel truck (Figure 6.2). These trucks serve basically as hose wagons and employ large reels for carrying large diameter hose. They may often carry 1 mile (approximately 1.6 km) or more of hose. The advantages of these reel trucks are that they can quickly lay out and pick up long distances of hose with a minimum number of personnel, and their ability to deliver large quantities of water over long distances with a single hoseline.

Figure 6.2 Some departments utilize special apparatus for laying long supply lines.

RELAY PROCEDURES

The traditional method of determining engine discharge pressure in relay operations employs the concept of setting up each pumping operation for the length of hose used and the amount of water that is ACTUALLY BEING PUMPED. It requires fireground hydraulic calculations such as EP = RP + TPL for in-line pumpers or EP = NP + PL for attack pumpers and requires refiguring every time another line is added, a nozzle flow setting is changed, or when additional length is added to an existing hoseline. This method frequently will not work on the fireground because of the following:

- Lack of training.
- Excitement, noise, and frequent interruptions affecting the pump operator's thinking process.
- Length of hoseline out is unknown.
- Use of multiple hoselines of different diameters makes figuring friction loss too complicated.
- Flow is unknown.
- Gauges are not correctly calibrated.

Determining the flow is particularly difficult because of the design of the two widely used types of fog nozzles: one type permits changes in flows by firefighters operating them (Figure 6.3);

Figure 6.3 An adjustable gallonage nozzle. *Courtesy of Elkhart Brass Mfg. Co.*

the other type automatically varies the flow according to the pressure (Figure 6.4).

A simpler and more efficient means of establishing a water relay is to use the constant pressure method.

EP = Engine pressure in psi or kPa
RP = Relay pressure of 20 psi (138 kPa) minimum for the next in-line pumper
NP = Nozzle pressure of attack line
TPL = Total pressure loss in psi or kPa due to friction loss and back pressure

The constant pressure relay depends on a consistent flow being provided on the fireground. The attack pumper can maintain this flow by using an open discharge or secured waste line to handle the excess beyond the flow being used in the attack lines. When fire incidents dictate the use of a relay, it is more common for the maximum relay flow to be committed when operations begin and then decreased as the operations progress.

Figure 6.4 An automatic nozzle. *Courtesy of Elkhart Brass Mfg. Co.*

Forming the Relay

Step 1: First-arriving engine is positioned at the fire. The person in charge of this company makes the size-up and determines the quantity of water needed. An initial attack is made with the water carried on the pumper.

Step 2: When the relay is being set up, the largest capacity pumper available should be positioned at the water source. In reality this is often not possible.

5-90

Step 3: In-line pumpers lay out their hose load (up to 1,000 feet [300 m]) (with two sections in reserve) for 2½-inch (65 mm) and 3-inch (77 mm) hose. Large diameter hose (LDH) relays can be longer, moving 1,000 gpm (3 785 L/min) with 2,000 feet (600 m) between pumpers.

Step 4: All supply lines are connected to the pumpers in the relay.

Step 5: The pump operator for each pumper, except the supply pumper, opens an unused discharge gate if the pump does not have a relay relief valve. This allows the air from the hoselines to escape as the water advances up the hoseline.

Step 6: The pumper at the water source pumps water at 175 psi (1 207 kPa).

Step 7: The pump operator at the next pumper closes the unused discharge gate once a steady stream of water flows from it, then advances the throttle until 175 psi (1 207 kPa) is developed. Each successive pump operator follows the same procedure.

Step 8: Each pump operator sets the pressure regulating devices.

Step 9: The attack pump operator adjusts the discharge pressure(s) to supply the attack line(s).

Step 10: The attack pumper maintains the flow during temporary shutdowns by utilizing one or more discharge gates as waste or dump lines (Figure 6.5). Attack lines should not be shut down except when absolutely neces-

Figure 6.5 The waste line should be directed away from traffic areas. *Courtesy of Bill Eckman.*

sary. If a hoseline bursts, a discharge gate of the in-line pumper before the rupture should be opened to dump water until the length is replaced.

Step 11: If additional water supplies are needed on the fireground, additional hoselines must be laid between the apparatus in the relay.

Pump operators should keep correcting to 175 psi (1 207 kPa) during the relay until:

- Intake pressure from pressurized sources drops to 20 psi (138 kPa). As the intake pressure drops below 20 psi (138 kPa) there is a danger that the pump will go into cavitation. Cavitation can be recognized by the fact that increasing the engine rpm's does not result in an increase in discharge pressure. Essentially, the relay capacity has been reached. The results of cavitation can be pump damage and/or disruption of the flow to the fireground.
- Operating the hand throttle does not result in an increase in rpm's. (Engine has reached governed speed.)

CAUTION: Due to the number of pumpers involved and the possibility of excessive pressure buildups, pump operators must remain alert during relay operations to compensate for pressure increases resulting from unanticipated situations (emergency shutdowns, rapidly closed nozzles, or equipment malfunctions).

A relay using constant pressure has several advantages:

- Provides a good fire flow.
- Speeds relay actuation.
- Requires no complicated calculations on immediate arrival.
- Lessens radio traffic and confusion between pump operators.
- Allows attack pumper engineer to govern fire lines with greater ease.
- Operators in the relay only have to guide and adjust pressure to one constant figure.
- Can be adjusted if necessary.

Flow for various size hose and layout conditions for an expected distance of 750 feet (229 m) between relay pumpers is given in Figure 6.6.

The constant pressure figure of 175 psi (1 207 kPa) can be modified as needed for

- Variations in relay pumper spacing
- Severe elevation differences between source and fire

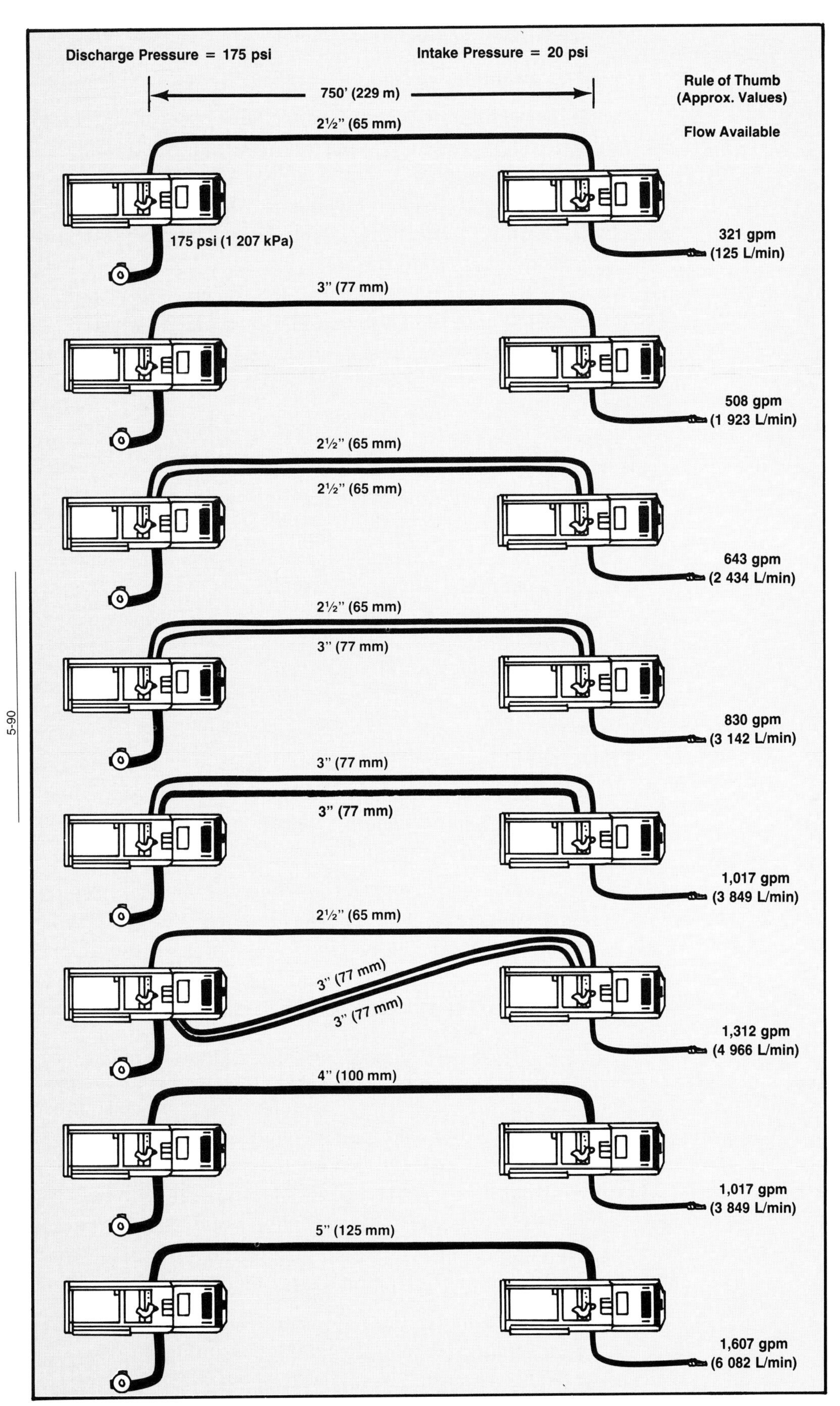

Figure 6.6 Shown are various flows for different size hose and layout conditions. All are based on an expected distance of 750 feet (225 m) between pumpers.

5-90

- Increases in needed fire flow
- Large diameter hose

When increasing the relay pressure, the supply pumper is adjusted until the operating pressure is reached. Each successive pump is similarly adjusted. When a decrease in the flow is required, the attack pumper throttles down. One way to do this is by opening the dump line to relieve excessive water. Successively, the in-line pumpers toward the water source throttle down to the new figure. The discharge pressure should not exceed 200 psi (1 379 kPa) if the hose has been tested at 250 psi (1 724 kPa). The water supply officer must realize the flow and pressure limitations of a given relay setup and should not attempt to exceed the capabilities of apparatus and hose.

Shutting Down

Pumps in relay are in effect a multiple-stage pump and enormous pressures can be built up. For this reason, do not close nozzles. Throttle the attack pumper down, but keep water discharging until the pump can be disengaged. Successively, the pumpers toward the water source follow the same procedure. An unused discharge gate is opened on each pumper. The water supply pumper is shut down.

HARDWARE

Protection of the pump and hoselines is a primary consideration when performing water supply relays. Appliances are available that will negate the need for discharge gates to be used as safety outlets and will simplify the job of the pump operators.

Relay relief valves are used to protect pumps and hoselines from excess flow and pressure. These valves are located on the intake side of the pump and can be set at the relay operating pressure. Pressure surges cause the valve to open and dump the excess flow harmlessly without creating changes in other operating units (Figure 6.7).

Bleeder valves are beneficial in relay operations (Figure 6.8). These valves prevent the large volumes of air present in uncharged hoselines from being driven into pumps and causing cavitation. They will also prevent air from entering attack lines and interrupting the water flow when the supply is switched from the pumper's tank to a relay. Bleeder valves may be kept open before pumping operations and should be kept open until a steady stream of water is flowing during operations.

Relays dependent on later arriving mutual aid companies can set up an initial relay of lesser volume and greater spacing with in-line relay valves placed in the relay line for the incoming pumpers. These valves allow the pumpers to hook up after a relay is operating and boost the pressure (and corresponding volume)

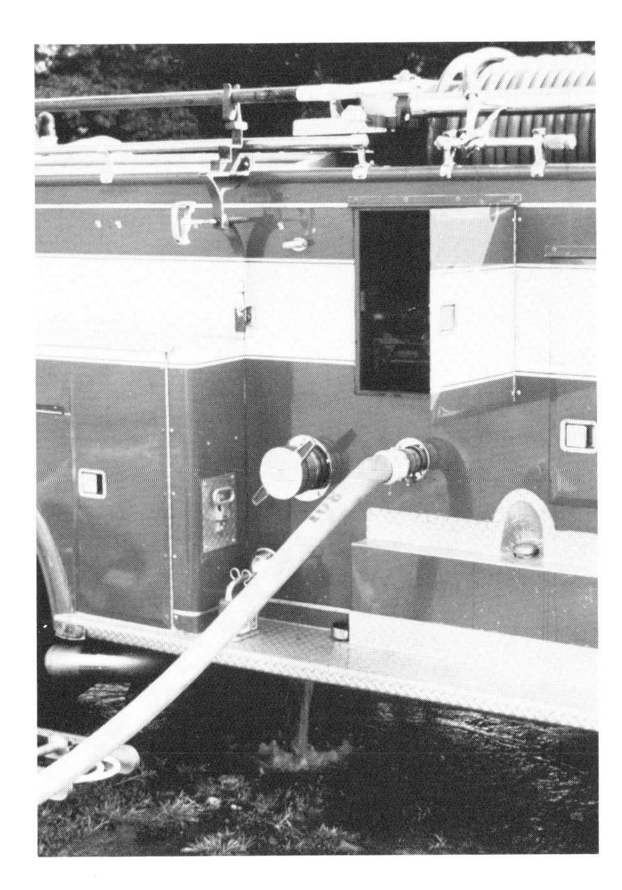

Figure 6.7 The intake relief valve will open at the pressure for which it is set. This prevents damage to the pump resulting from excess pressures. *Courtesy of Bill Eckman.*

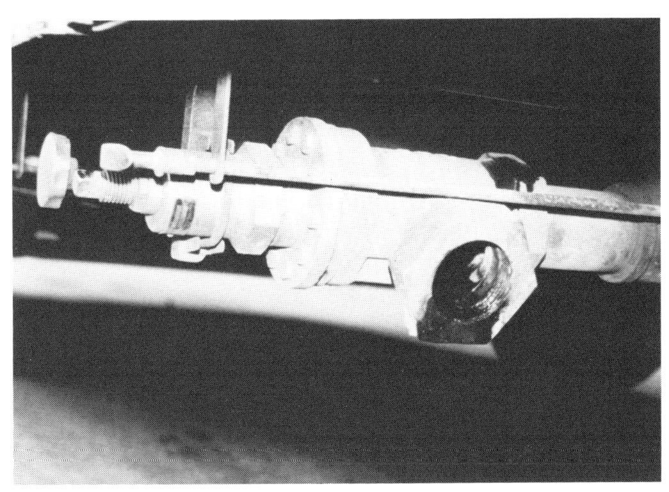

Figure 6.8 View of a bleeder valve.

without interrupting operations (Figure 6.9). Fire departments that do not have a valve of this type available can construct one simply by the use of a 2½ x 2½ inch (65 mm by 65 mm) gated wye and a 2½-inch (65 mm) clappered siamese (Figure 6.10).

Figure 6.9 An in-line relay valve can be placed into a supply line before the line is charged; then, later-arriving apparatus can enter into the relay to boost pressures. *Courtesy of Jaffrey Fire Protection Co., Inc.*

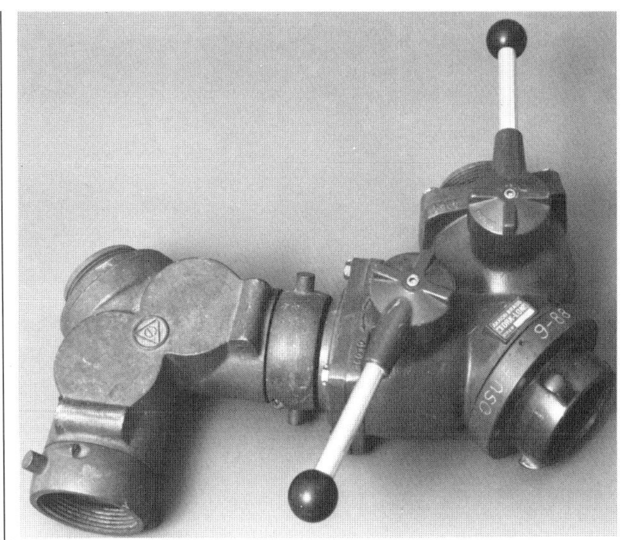

Figure 6.10 An in-line relay valve can be constructed from a gated wye and a clappered siamese.

SUPPLEMENTAL PUMPING

The term "supplemental pumping" can be misleading in that it is used to describe several different types of operations. However, all these different operations have the same purpose: to provide additional water to a pumper when the hydrant it is already attached to is not supplying enough. This may become necessary when the hydrant is unable to supply the amount of water that the pump is rated for, or when the intake hose already in use is not capable of flowing the rated capacity of the pump or the water available at the hydrant.

The most common type of supplemental pumping is used when a pumper is unable to obtain the maximum amount of water available due to the intake hose. In this case it may be desirable to add one or two 2½- or 3-inch (65 mm or 77 mm) lines between the hydrant and the pump. If a hydrant gate valve were placed on the hydrant's 2½-inch (65 mm) connections prior to turning the hydrant on the first time, this would be a simple task. Just connect the line(s) between the two and open the gate valves. However, if no gate valves were attached, it would be necessary to shut the hydrant down to make the connections. Since the flow of water through discharge lines will be interrupted, attack crews should be warned first. If the decision is made to shut down the hydrant, the hose should be hooked to the pumper before the hydrant is shut down. This minimizes the amount of time the water supply is interrupted. Often, rather than shut the hydrant down, it is desirable to run supplemental lines to another hydrant and either operate directly off the second hydrant or have another pumper supply the line from the second hydrant.

In extreme cases, water can be pumped from a hydrant on a strong water main into a hydrant on a weaker main in order to boost the available water in the weaker main. This is not a common operation and is generally discouraged by water officials, due to its potential for contaminating the supply system. Figure 6.11 shows several different layouts for supplemental water supply operations.

Water supply officers can utilize designated water supply companies to establish supplemental flow from different hydrants, from larger and/or stronger mains, or from static sources. Ingenuity must be used to determine means of utilizing all the available water in mains and other sources. Tanker shuttles and relays can be used as supplemental supply. Using the term "low water pressure" to rationalize large fire losses is only true in a few cases. Often, it is the poor utilization of the available water or failure to identify additional water resources that must be critically reviewed. Proper planning and training in water supply operations must be performed so the maximum flow can be provided for any incident that may occur.

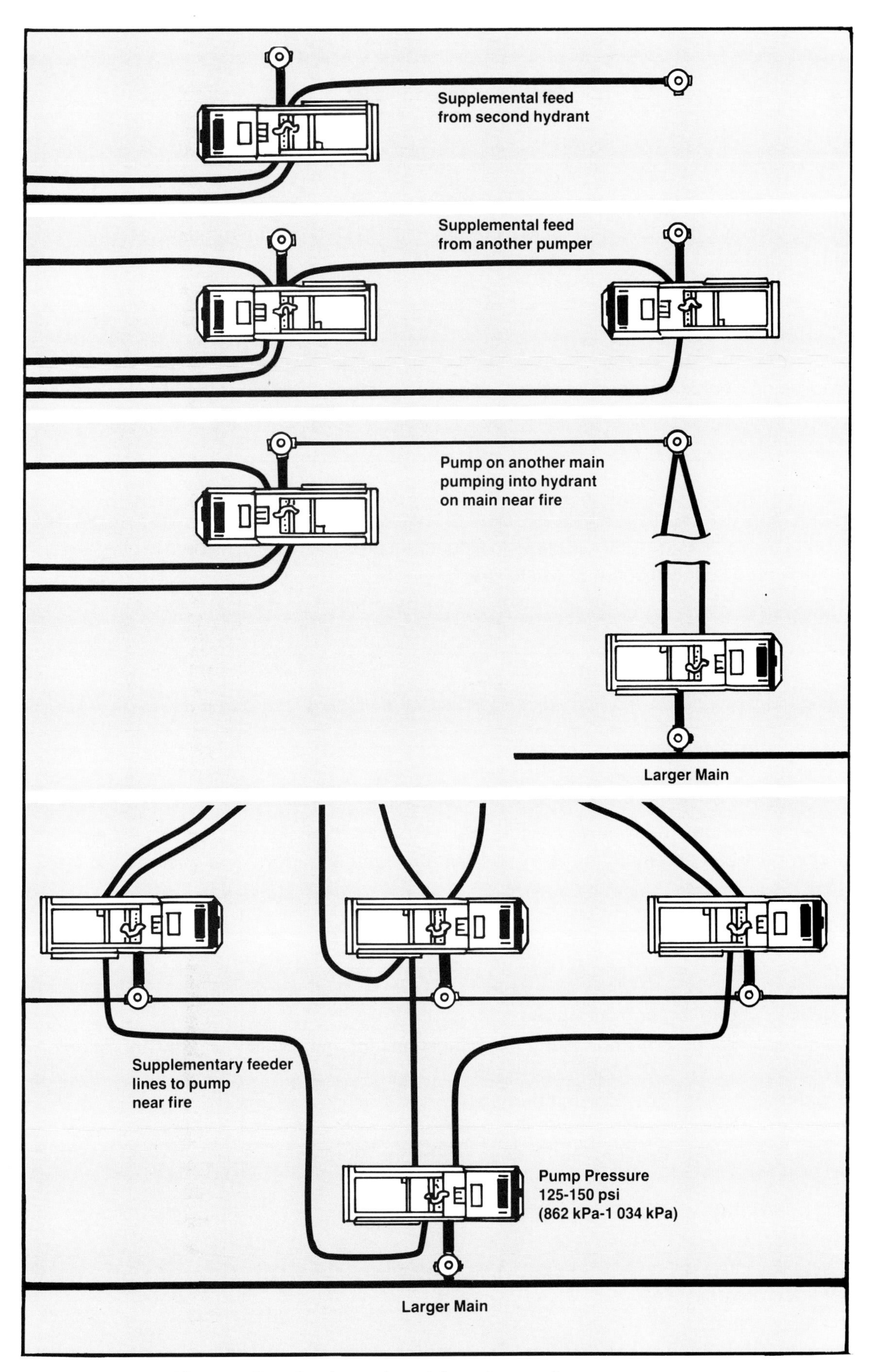

Figure 6.11 Several different configurations for supplemental pumping operations.

Chapter 6 Review

Answers on page 262

TRUE-FALSE. Mark each statement true or false. If false, explain why.

1. When using a constant pressure relay, it is more common to start a minimum flow and increase the flow as needed.

   ☐ T ☐ F ______________________________

   ____

2. When establishing a constant pressure relay, the source pumper determines the quantity of water needed.

   ☐ T ☐ F ______________________________

   ____

3. To handle excess water at the attack pumper in a constant pressure relay, it is best to open an unused discharge or to supply a secured waste line.

   ☐ T ☐ F ______________________________

   ____

4. When raising relay pressures, the first adjustment is made at the attack pumper.

   ☐ T ☐ F ______________________________

   ____

5. The advantages of using a relay valve are that it allows water to be flowed immediately and that additional pumpers can enter into the relay without having to shut down the operation to make connections.

   ☐ T ☐ F ______________________________

   ____

MULTIPLE CHOICE: Circle the correct answer.

6. Which one of the following is *not* one of the basic functions for apparatus in a relay operation?
   A. In-line pumper
   B. Attack pumper
   C. Source pumper
   D. Support pumper

7. During a relay, the starting pressure for the source pumper should be ________ psi (kPa).
   A. 100 psi (690 kPa)
   B. 150 psi (1 034 kPa)
   C. 175 psi (1 207 kPa)
   D. 200 psi (1 380 kPa)

8. Which of the following is *not* an advantage of a constant pressure relay?
   A. Greater flows are possible
   B. Speeds relay actuation
   C. Lessens radio traffic and confusion
   D. Requires no complicated calculations

9. What is the primary hazard involved in supplemental pumping operations?
   A. Main or pipe collapse
   B. Contamination of the water supply system
   C. Apparatus pump damage
   D. High pressure in hoselines

SHORT ANSWER: Answer each item briefly.

10. Name at least two factors that can limit a relay operation.

____

____

11. Traditional formulas, such as EP = RP + TPL, used to determine relay or engine discharge pressures, often do not work in fireground situations. Why?

____

____

12. You are a pump operator at an in-line pumper in a constant pressure relay. The pumper is not equipped with a relay relief valve. What is the best way to discharge the air from the incoming hoseline?

____

____

13. Name the two situations during which the discharge pressure can no longer be adjusted to 175 psi (1 207 kPa).
    A. ____
    B. ____

14. Describe how to construct a makeshift relay valve.

____

____

15. What is the purpose of supplemental pumping?

____

____

16. Name two functions that must be performed to ensure that a maximum water flow can be achieved under fire scene conditions.
    A. ____
    B. ____

Photo by Ziamatic Corp.

Chapter
7

Shuttle Operations

This chapter provides information that addresses performance objectives in NFPA 1001, *Fire Fighter Professional Qualifications* (1987) and NFPA 1002, *Fire Apparatus Driver/Operator Professional Qualifications* (1988), particularly those referenced in the following sections:

NFPA 1001

Fire Fighter II

4-15.7

4-15.8

4-15.9

NFPA 1002

3-2.12

Chapter 7
Shuttle Operations

APPARATUS

In shuttle operations, two types of apparatus will be required in the water supply system: pumpers and tankers (also known as tenders). Pumpers can be used as follows: (Figure 7.1)

- At the source to fill tankers after they have unloaded
- To unload tankers that have inadequate pump capacity and/or no dump valve
- To draft from a portable tank or nurse tanker at the unloading site to supply fireground apparatus

Figure 7.1 Several types of pumpers. *Courtesy of Joel Woods.*

- As a shuttle tanker (pumper-tanker)
- As the attack pumper

Pumpers serving on the fireground, called attack pumpers, carry on normal fire suppression activities. The attack pumper engineer must realize that the number and flow of attack lines is governed by the available water supply. Beyond this consideration, the engineer should not have to be concerned with the supply operation itself. The water supply officer should institute procedures to maintain an adequate water supply.

Mobile water supply apparatus, tenders, or "tankers," are widely used to transport water to areas beyond the water system or areas where water supplies are inadequate and must be supplemented. Although attack pumpers carry water, most do not carry enough water to be used effectively as tankers unless they are of the pumper-tanker category (Figure 7.2).

Courtesy of Joel Woods.

Courtesy of Joel Woods.

Courtesy of Bob Esposito.

Figure 7.2 Several pumper-tankers.

The size of the tank will depend on the department's water requirements and the weight and capability of the apparatus. If an approved tanker is desired, the requirements of NFPA 1901, *Standard on Automotive Fire Apparatus,* should be met. The road test and weight distribution requirements generally limit tank capacity to less than 1,500 gallons (5 678 L) on a single axle vehicle (Figure 7.3). Where tanks of 1,500 gallon (5 678 L) capacity or larger are used, tandem rear axles or semitrailer construction, or both, may be needed (Figure 7.4). Nonstandard tankers may be as

Figure 7.3 These smaller tankers (under 1,500 gallons [5 678 L]) require only one rear axle, thus giving them excellent maneuverability. *Courtesy of Bob Esposito.*

Figure 7.4 Larger tankers require at least two rear axles. *Courtesy of Joel Woods.*

large as 2,500 gallons (9 464 L) on a single axle and 8,000 gallons (30 283 L) on a semitrailer (Figures 7.5a and b).

Figure 7.5a Semitrailer tankers may carry up to 8,000 gallons (30 283 L) of water. *Courtesy of Joel Woods.*

Figure 7.5b Areas with a severe lack of water supply may require a large amount of "water on wheels." *Courtesy of Bob Esposito.*

Tankers over 4,000 gallons (15 142 L) provide the advantage of being "a rolling water source." One large tanker may provide more than enough water for fire control on residential and medium-size structures. These large tankers function well as stationary fill points and nurse tankers for attack pumpers or can be used as reservoirs by shuttling tankers. A nurse tanker serves as a large volume storage point and may provide a greater water supply safety factor for the attack pumper than a single portable tank due to its size (and in some cases pumping capability).

The problems with large tankers are that they may be over bridge and road weight limits, have a slower safe speed, and have trouble traversing narrow bridges and roads, increasing response time. When fire incidents require long duration water supply, tankers carrying greater than 4,000 gallons (15 142 L) are best utilized as a nurse tanker or reservoir. Medium-size tankers (1,500 to 3,500 gpm [5 678 L/min to 13 249 L/min]) will generally have the best gallon per minute (L/min) capability and will be more efficient for hauling water.

Pumper-tanker combinations are apparatus that can be utilized as either a tanker or a pumper, depending on the need (Figure 7.6). These vehicles should have at least a 1,000 gallon (3 785

Figure 7.6 Pumpers with water tanks of 1,000 gallons (3 785 L) or larger may also be used as tankers if needed. *Courtesy of Bob Esposito.*

L) tank capacity and a recommended pump capacity of 1,000 gpm (3 785 L/min). Vehicles of this type may also be used as "attack tankers" depending on their primary response function.

Fire departments that cannot afford to purchase manufactured tankers can convert other types of tank trucks to fire service use. Tank trucks that may be converted for water supply include petroleum tankers, milk trucks, and vacuum trucks (Figure 7.7). This is a viable option to the purchase of manufactured pieces, but requires the review of several critical criteria:

- Previous product hauled, its density, and effect on Manufacturer's Gross Vehicle Weight Rating
- Contamination from previous use
- Conversion effect on acceleration, braking, handling, and safety due to baffling and tank security

Water has 10 percent greater weight by volume than fuel oil. A fuel oil tanker of 1,000 gallons (3 785 L) capacity filled with water would create an additional 800 plus pounds (363 kg) that the vehicle was not designed for. A gasoline tank truck would have an even larger difference since it is 20 percent lighter in weight than water.

Figure 7.7 Many departments have successfully converted old milk or oil tankers into fire fighting apparatus. *Courtesy of Joel Woods.*

Fuel tankers are also designed for filling at the start of the day and being unloaded as they complete distribution runs. Vehicles of this type do not generally sit "under load" for extended periods of time. Problems that can occur from this extra weight are chassis damage, braking inability, tire damage, suspension damage, poor handling characteristics, and insufficient engine power. Tank capacity may need to be limited or truck components upgraded to eliminate these problems with converted vehicles. The tank to frame mounting generally requires strengthening to support the excess weight of water over fuel oil or gasoline.

Piping will need to be added for filling and unloading operations and pumps will require upgrading for fire department use (Figure 7.8). Ordinary high pressure fuel oil pumps do not deliver adequate capacity for fire fighting purposes. Quite frequently, commercial tankers will not have baffles. Due to reduced handling abilities, driving a half-loaded tanker without baffles would be very hazardous. It is a good practice to drive any unbaffled tankers either fully loaded or completely empty.

Figure 7.8 When converting a vehicle into a mobile water supply apparatus, consider expanding the fill and discharge lines. Above, a second line has been added to the existing line.

Regardless of the origin or nature of a department's tanker, the maximum size and weight of any apparatus is limited by the individual response area. Narrow winding roads, hills, small rural bridges, load-bearing capacity of roads and bridges, and rough terrain must all be taken into consideration when purchasing a tanker.

All tankers should conform to the following construction requirements in order to be both safe and efficient water movers:

- Adequate but reasonable tank capacity (1,000 gallons or greater [3 785 L or greater]).

- Adequate loading rate (500 gpm or greater [1 893 L/min or greater]).
- Adequate unloading rate (500 gpm or greater [1 893 L/min or greater]).
- Adequate vent capacity (Figure 7.9).
- Observe Manufacturer's Gross Vehicle Weight Rating.
- Good roadability
  — Adequate chassis
  — Adequate power
  — Adequate brakes
  — Proper tank mounting
  — Proper baffling

Figure 7.9 The holes in this tank vent cover are too small to ensure maximum loading or unloading of the tank. This lid should be fully opened during loading and unloading. *Courtesy of Bob Esposito.*

HARDWARE/EQUIPMENT

Portable Pumps

Portable pumps are often standard accessories on urban and rural fire apparatus. They can serve many functions when a pumper is unavailable or unable to do a task. Portable pumps can draft in locations that are inaccessible because of distance or terrain. Floating pumps can be used as the water source to supply pumpers or for attack lines for nearby fires.

The major factor to be considered when portable pumps are used to supply water for fire fighting is their pumping capacity. Most small portable pumps can only supply 250 to 300 gpm (946 L/min to 1 136 L/min). Water relays or tanker filling operations will not work effectively with such limited flows. Several pumps may need to work in parallel from the same source to maintain an adequate fill site water supply. Some departments have used a large volume low pressure floating pump to supply a fill site pumper (Figure 7.10). Larger portable pumps are sometimes mounted on trailers and hauled to the drafting site. They provide the advantage of freeing pumpers for fireground or relay operations. However, pumpers at the drafting site also have advantages; they are generally more reliable and provide signaling, communications, and a point for basing operations.

Figure 7.10 High volume, low pressure floating pumps are excellent for use in filling tankers. *Courtesy of Bill Eckman.*

Portable Tanks

Portable water tanks are highly beneficial to water supply operations and may serve many functions. Commercially available tanks generally range in size up to 3,000 gallons (11 356 L) (Figures 7.11a and b), while locally built tanks have been constructed up to 6,000 gallons (22 712 L) (Figure 7.12).

Figure 7.11a Portable tanks may be carried in the same manner as a ladder. *Courtesy of Bob Esposito.*

Figure 7.11b Folding portable tanks are put into service by simply pulling on each side of the tank. *Courtesy of Bob Esposito.*

Figure 7.12 Many departments have built their own folding tanks.

There are several types of portable tanks. The most common is the collapsible or folding style using a rectangular metal frame and canvas duck or synthetic liner. This unit can be carried by two firefighters and set up in a matter of seconds, preferably on a flat level area with no sharp objects under the tank that could puncture the liner. These tanks can be carried in hose bed compartments, hinged racks above the hose bed, enclosed compartments, or racks on the side of pumpers or tankers (Figure 7.13).

Figure 7.13 Folding tanks may be stored in the side of the apparatus in a manner similar to ground ladders. *Courtesy of Bob Esposito.*

Another style is a round synthetic tank with a floating collar that rises as the tank is filled, making it self-supporting (Figure 7.14). This device weighs 35 pounds (16 kg) and merely requires unrolling. It can be set on uneven ground, and inlet-outlet connections are arranged in the bottom of the tank. This unit folds to approximately duffle bag size, can be carried preconnected, and fits on a pumper or tanker. A third unit is supported by four metal corner supports. This tank folds to a size similar to a hard suction hose and can be carried in a trough or compartment on a pumper or tanker. This unit tends to collapse when a hard suction hose is placed over the unsupported wall of the tank. The last unit is a locally constructed tank made of folding components that are assembled to build the portable basin. These take longer to set up, but can be built in a desirable 6 x 12 foot (1.8 m by 3.7 m) configuration that only blocks one lane of traffic. Since several of these interchangeable parts can be used together, portable tanks can be assembled in larger sizes than those commercially available. These "home-built" tanks may be somewhat larger and require more space on a tanker.

Figure 7.14 Floating collar tanks erect themselves as water is added to them.

The main functions of portable water tanks are to serve as a reservoir for attack pumpers and to provide an unloading point for tankers (Figure 7.15). Tanks should be located to allow several tankers access at one time and to ease traffic congestion at the unloading site. Pumpers should draft in one of the following ways:

- From the corner of the tank, with the hard suction hose laying across the longest portion of the tank (Figure 7.16)
- From the side of the tank, with the hard suction laying across the middle of the tank (Figure 7.17)

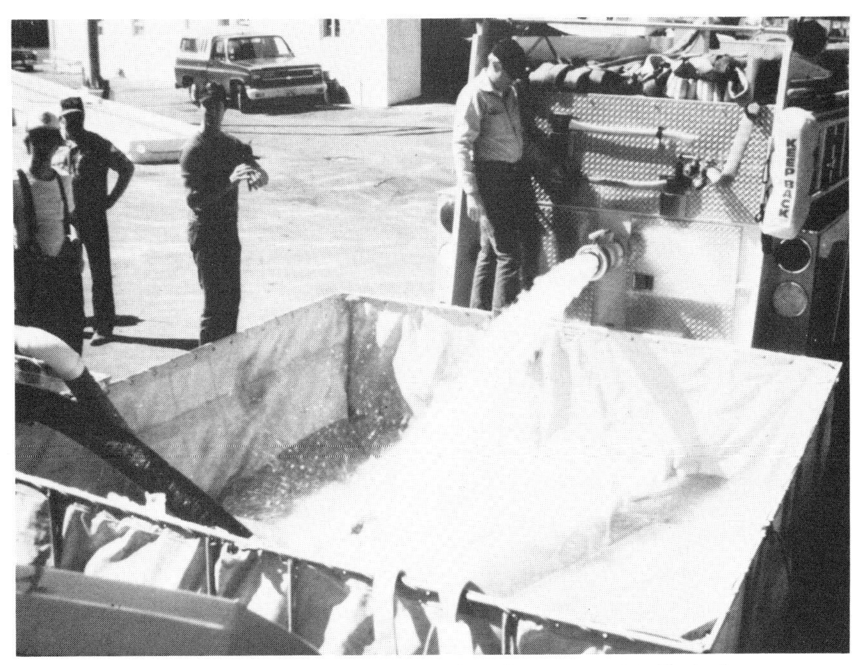

Figure 7.15 Apparatus should be positioned so that all water goes into the tank.

Figure 7.16 By drafting out of the corner of the tank, the driver/operator has more room to work at the pump panel. *Courtesy of Bob Esposito.*

Figure 7.17 By drafting from the side of the portable tank, the driver/operator has less room to move around the pump panel. *Courtesy of Bob Esposito.*

Special strainer equipped suction devices make it possible to use water down to 1 or 2 inches (25 mm or 51 mm) from the bottom of the reservoir (Figures 7.18a and b).

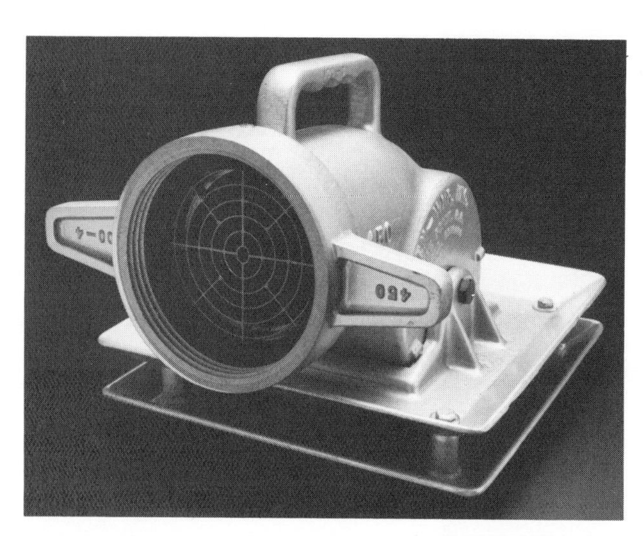

Figure 7.18a This commercially built strainer is designed for use in portable tanks and can take in water at depths down to 2 inches (51 mm) at flow rates up to 1,000 gpm (3 785 L/min). *Courtesy of Ken-Mar, Inc.*

Figure 7.18b This homemade portable tank strainer will also draft water to a depth of 2 inches (51 mm). It is made of PVC pipe and reduces the problem of getting the hard suction hose to lay flat in the bottom of the tank. *Courtesy of Bob Esposito.*

Siphons/Tank Connectors

Frequently, one tank will not hold enough water for maintaining effective fireground flows. In these situations, multiple tanks must be used. The suction hose should not be changed from tank to tank during pumping operations unless absolutely necessary. If it must be moved, the intake valve, if available, should be closed and the hose transferred, rather than shutting down the pump and having to reprime. Siphons or tank connectors may be used to avoid this situation and provide tank water level continuity. There are several types that serve basically the same function (Figures 7.19a-c).

Figure 7.19a This hard pipe connector allows two tanks to be used at the same time.

Figure 7.19b Tanks may also be connected with a clamp that connects the two tank drains. *Courtesy of Marv Ackerman.*

Figure 7.19c Two tanks connected by drains. *Courtesy of Rich Mahaney.*

Jet siphons may be used to maintain the water level in the drafting tank while tankers dump into others. The siphon utilizes a 1½-inch (38 mm) discharge line to a jet in the siphon set between two tanks (Figures 7.20a-c). The attack pumper will draft from the reservoir into which the siphon is dumping. There are several methods of utilizing this type of venturi jet, not only for

Figure 7.20a A commercially built jet siphon device. *Courtesy of Ziamatic Corp.*

Figure 7.20b A homemade jet siphon device.

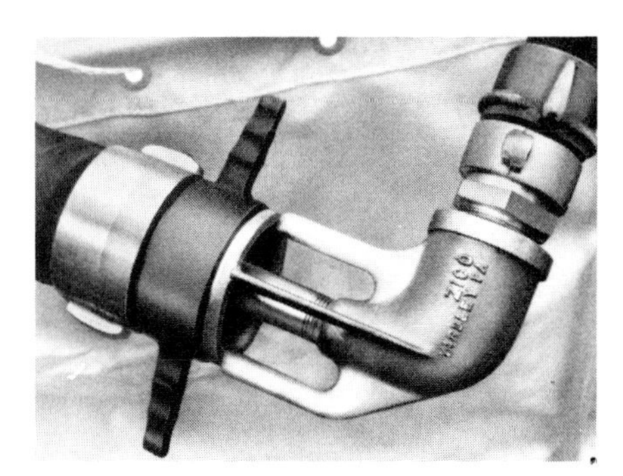

Figure 7.20c A jet siphon as it would be attached to a hard suction hose. *Courtesy of Ziamatic Corp.*

water transfer between tanks but also to speed filling and unloading operations. Jet siphons allow the attack (or supply) pumper to maintain an adequate water supply from the secondary tanks in the primary drafting tank. A minor drawback of this method is that an additional discharge line must be set up and controlled by the attack (or supply) pumper engineer. The jet siphon may be either a hard suction hose or PVC tube designed to go between the tanks.

Another means of transferring water from one tank to another is through the use of siphon piping. Water will flow to maintain an equal level between the two tanks as long as the siphon is properly primed. The use of these tubes effectively turns two (or more) portable tanks into one tank with a capacity equal to the combination of the single tanks. The siphon tubing must be as large in diameter as practicality allows. Small tubes are unable to maintain a flow rate that will keep up with a large pump discharge. If fire flows in excess of 500 gpm (1 893 L/min) will be used, a jet will be needed in the siphon base (Figure 7.21). Multiple siphon tubes will also aid the flow capacity. Storage space on the tanker or pumper will be required for the multiple tubes. Positive flow control between tanks is more desirable for shuttle operations and should be maintained whenever possible. Remember a straight siphon will only keep the tanks level, while a jet can use most of the water out of a tank.

Figure 7.21 This hard pipe tank connector incorporates a jet stream to speed the transfer from one tank to the next.

Another means of transferring tank water is through a more permanent tank modification. A short piece of hard suction or large diameter hose is fixed to the sides of two (or more) portable tanks at their base. This will maintain a gravity flow effect until the water equalizes between the tanks. An arrangement of the tanks in this manner will require substantially more space to accommodate both tanks on the fireground. Tanks may also be obtained with a drain tunnel connector that allows two or more tanks to be connected with a clamp on the drains or by inserting a piece of PVC pipe in the drains. This arrangement allows separate, single tank storage.

These siphons and connections increase water tank capacity, provide a water supply safety factor for the attack pumper engineer, and allow rapid unloading for tanker shuttle operations.

Portable water tanks can serve other functions. These may be used to provide a safety valve in an open relay. In this function, they allow water surges to be dumped harmlessly into the tank instead of causing a water hammer in the next in-line pumper (Figure 7.22). They create a floating reservoir that assists in maintaining the water supply during changes in demand. Portable tanks may serve in a similar fashion when hose thread differences occur between two mutual aid departments. Hydrants with different or damaged threads may be discharged into a portable tank, and then the water drafted by the pumper. Low flow hydrants can also be discharged into a portable tank. The supply pumper can then gauge flow to meet the demand at the peak periods and use the slack time to allow the tank to refill, or augment the supply with tankers dumping into the portable tank.

Figure 7.22 This portable tank is being used to catch water in an open-end relay. This type of relay reduces the chance of dangerous pressure being built by pumps in stage.

DELIVERY METHODS

Tanker shuttle operations are quite often the only means of delivering water to rural fire incidents. Shuttles can also be used to supplement large diameter hose systems and relays, or be used while the relay is being established.

Tanker shuttle operations are conducted in a wide variety of methods. The procedures used by a department depend on the availability of support apparatus (pumpers, devices, fittings, adapters, portable tanks) and the design and distribution of tankers (nurse and shuttle).

Pumping Procedures

Efficient pumping procedures at the fill site can save valuable time. Pumpers should receive the selected fill site location from the water supply officer (WSO) or their water source maps while en route. Upon arrival, the officer must determine the best position for drafting that will also allow maximum access for the later arriving tankers. Ideally, the engineer's panel should be located to view both the source and fill operations, but generally will have to be away from the water with the intake hose on the opposite side allowing the engineer to view the tanker filling operations.

Once the pumper is at the water source, stop the apparatus near the drafting location and connect the hard suction to the intake. Connecting the hard suction before moving to the final drafting point will help avoid the following:

- Unsure footing for firefighters connecting hose while standing in water with tides or currents
- The possibility of dropping the hard suction over seawalls or embankments and losing time while retrieving the hose or possibly damaging it

Once the suction and pumper are positioned, the drafting operation should be established according to the steps outlined in IFSTA's **Operation of Fire Department Pumping Apparatus**. When necessary and possible in tanker fill operations, the throttle should be increased until the increase in rpm results in no increase in discharge pressure. This point is the maximum capacity of the pump. Once this is reached, the throttle should be backed down to allow for a safety factor. The pump will now be running near cavitation and requires extra caution. When a pumper is filling a tanker or portable tank with an open end connection, considerably less pressure may be used. This is because personnel will usually be manually holding the end of the hose and excess pressures could result in harm to them. When open end connections are necessary, supply only the maximum pressure that can be safely handled by the personnel holding the end of the hose.

Factors that may require a decrease in the fill rate include the following:

- Number and size of available lines
- Line reaction when filling overhead

- Inadequate tank venting
- Improper baffling
- Inadequate source

Once a pumper has started discharging in a tanker filling operation, it should not shut down and risk the possible loss of prime. Additional time could be lost if the pump required priming before filling each arriving tanker. To avoid losing prime, an additional discharge line is stretched to the water source. This line is operated as needed between filling tankers to eliminate shutting down the pump. Another method to achieve this is to flow a booster line constantly back to the source. This will not seriously affect the flow rate and will both maintain continuous operation and keep the pump cool during the period between filling tankers.

All tanker fill lines should have a positive flow control device near the terminus. Gated valves, wyes, water thiefs, or hose clamps can be used (Figures 7.23a and b). The device should be placed close to the make and break position so personnel can make the hoseline connection and immediately initiate the flow. If sufficient control devices are not available, discharge gates on the pumper must be used to control the water flow to the tankers.

Figure 7.23b Gate valves on hydrants may also be used to control fill lines. *Courtesy of Rich Mahaney.*

Figure 7.23a Hose clamps may be used to control tanker fill lines.

Tanker Fill Method

Once fill pumpers are in place and prepared to operate, a number of methods are available to deliver the water to the tankers:

- Top fill openings (Figure 7.24)
- Direct tank inlets, gated and ungated (Figure 7.25)
- Large tank discharges (dump valves) that are threaded or have a siamese (Figure 7.26)
- Pump to tank piping — Tank fill line
- Tank to pump piping — Tank suction line (when unchecked)

Figure 7.24 Tank vents may also be used to fill tankers. *Courtesy of Bob Esposito.*

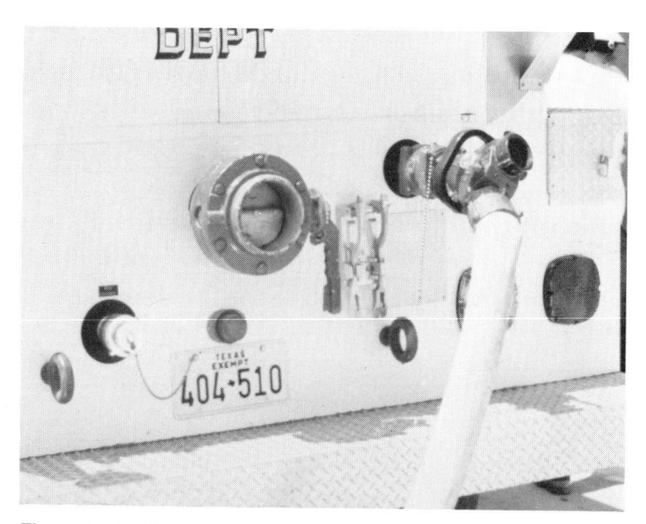

Figure 7.25 Some apparatus are equipped with direct tank inlets to speed filling.

Figure 7.26 Threaded rear discharges may also be used for filling operations. A siamese may be added when using multiple small diameter lines.

Overhead pipes for top fillings may be either permanent or portable. Portable fill pipes are generally made of PVC or other lightweight material and are carried on the apparatus for use at any water source. One method of operation is to place the end of the fill pipe in a static source, and pump into connected in-line jets (Figure 7.27) to generate high flow rates. These in-line jets should be pumped at 150 psi (1 034 kPa) and will provide 700 to 800 gpm (2 650 L/min to 3 028 L/min) through 4-inch (100 mm) pipe. Larger diameter piping will provide additional flow.

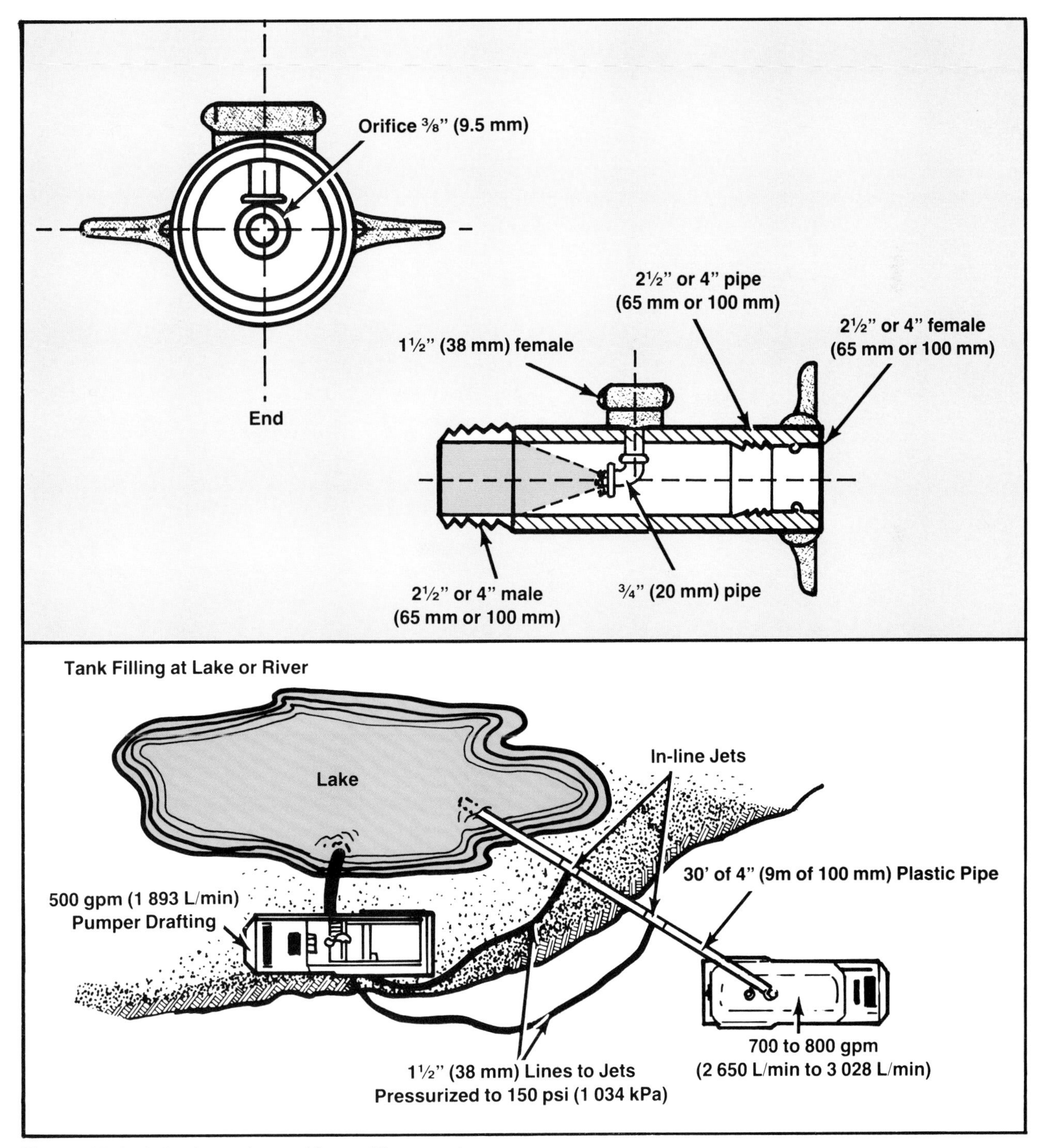

Figure 7.27 An in-line jet can be used for top filling operations.

Figure 7.28 Permanent fill pipes, to be supplied by pumpers, can be installed at locations with dependable water supplies. *Courtesy of Jim Berggren.*

Another method of overhead pipes uses permanent or portable manifolds. The permanent type are located adjacent to a water source and are fed by a fill site pumper(s) (Figure 7.28). These require tankers with large top fill openings in order to get maximum flow. Water sources may be hydrants, irrigation pumps, or static supplies. The portable type may be lightweight piping carried on a unit to the fill site and then hooked over the top opening (Figure 7.29) or a multi-inlet manifold that is elevated above the top opening by outside means (Figure 7.30). Due to the number of lines feeding these devices and their diameter, the flow capacities can vary from 600 to 1,000 gpm (2 271 L/min to over 3 785 L/min).

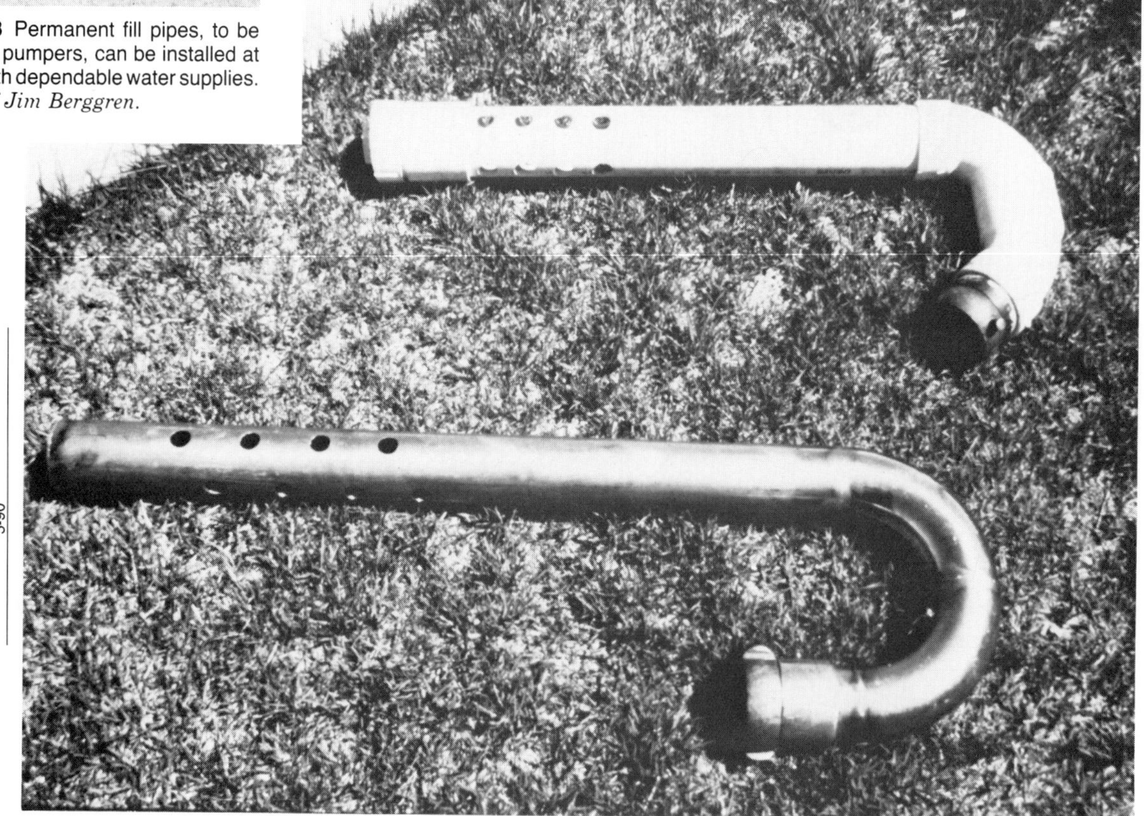

Figure 7.29 Portable fill pipes can be made to hook over the fill opening of the tanker.

Figure 7.30 This temporary fill pipe is supplied by an irrigation pump and is supported by outside means.

Filling a tanker through the top without using a fill device is not recommended due to reaction of the hoseline. Firefighter safety is a primary concern, and firefighters can be thrown off or slip from a tanker during top filling. There are several fill devices that can be utilized to fill a tanker through the top (Figure 7.31). These may include a variety of stream diffusers, both locally built or available in the department, to reduce the pressure reaction of the fill line.

Tankers may be filled by running discharge lines directly from the fill pumper or hydrant. Almost any larger line including hard suction hose, 2½-, 3-inch (65 mm, 77 mm), or large diameter hose can be utilized as fill lines.

Direct tank inlets are usually 2½-inch (65 mm) gated lines into the tank. Many tankers have been built with direct piping that extends to the top of the tank to eliminate the need for valving. Normally, two lines, often 3-inch (77 mm), are used (Figure 7.32). Newly designed units feature 4- or 5-inch (100 mm or 125

Figure 7.31 Stream diffusers direct water out the sides of the pipe, thus reducing the back pressure on the person holding the hoseline in the fill opening.

Figure 7.32 This tanker has two 3-inch (77 mm) permanently mounted fill pipes on the back of the tank.

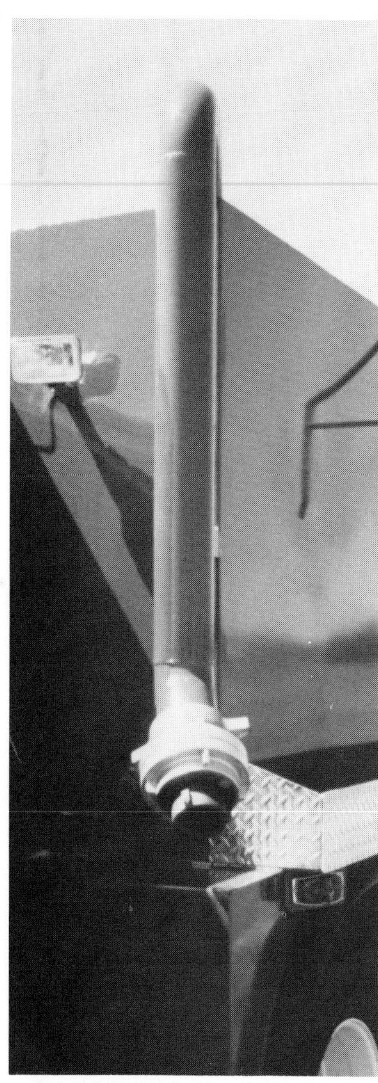

Figure 7.33 One large diameter fill pipe may be permanently mounted to the apparatus to speed filling. Quick-connect couplings will speed the making of connections.

mm) pipe and quarter turn or cam lock couplings to increase the flow and decrease make and break time (Figure 7.33).

Tankers may also be filled through a threaded tank discharge (dump valve). The size hose used will depend on the diameter of the discharge or the use of a gated suction siamese (Figure 7.34). Due to the size of the opening and valve, this is generally the fastest way to fill a tanker.

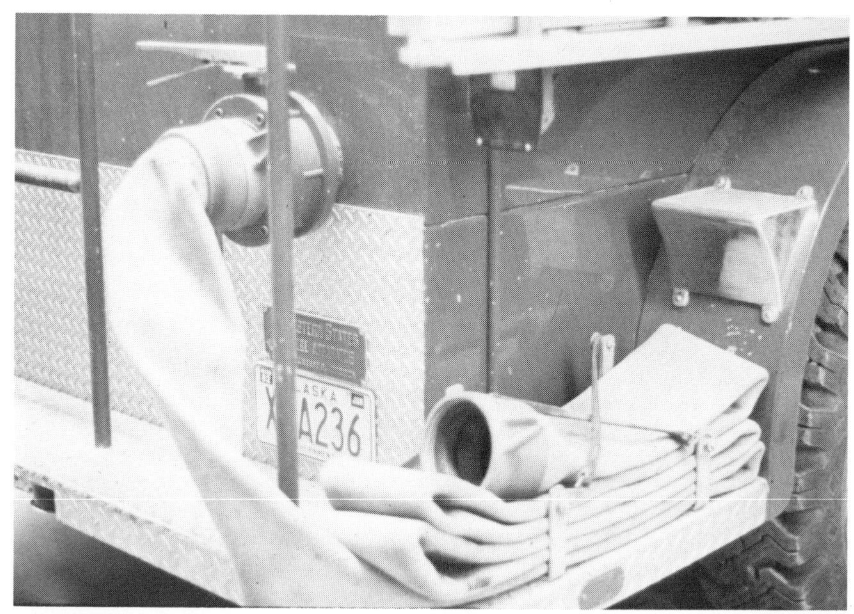

Figure 7.34 This tanker has a section of large diameter hose preconnected for maximum speed in filling operations.

When taking water into the pump inlet on a tanker or tanker pumper, there are two ways of getting it into the tank. First, the pump can be engaged and the tank fill line used; however, this is normally a small lower capacity line and high pressure is not desirable. Second, high flows to the tank can be obtained by opening the tank suction(s) and allowing water to back feed into the tank. Some pump models have a check valve that prevents flow from the pump to the tank through the tank suction line(s). Thus, the check valve acts to prevent overpressurizing closed tanks. The fire department should check with the manufacturers to determine if this can be done, and if adequate tank vents are available.

Regardless of the method used, all tanker operations require sufficient vent capacity to allow the air to escape the tank at larger fill rates. Inadequate venting can cause bulging and possible rupture of tankers when filling at high volumes (Figure 7.35).

Although sometimes difficult to maneuver, larger diameter lines will reduce fill times and speed the overall fill site operation. When 2½- or 3-inch (65 mm or 77 mm) lines are used, multiple lines will be needed to maintain high flow rates; with 4- or 5-inch (100 mm or 125 mm) fill lines a single line will move sufficient quantities (Figure 7.36). When large fill site operations are required, loading operations can be augmented by using two pumpers. This action cuts fill times in half since two or more units are filled simultaneously. In some cases of well-vented large capacity tractor-trailer units, two pumpers may be used to fill the same tanker.

Figure 7.35 Although the tank pictured here is a milk truck, it emphasizes the amount of damage that results from insufficient or closed tank vents during unloading operations. *Courtesy of David Grupp.*

Figure 7.36 Large diameter hose speeds tanker fill times.

Special source pumpers can be used for filling operations. These consist of small units equipped with large volume auxiliary powered irrigation or trash pumps discharging through large diameter hose. Source pumpers have capacities to 1,600 gpm (6 057 L/min) at a maximum of 80 psi (550 kPa). With properly designed tankers, fill rates in excess of 1,500 gpm (5 678 L/min) are not unusual (Figure 7.37).

Figure 7.37 Departments that perform a large number of tanker fill operations have special apparatus designed to operate at fill sites. *Courtesy of Rich Mahaney.*

Tanker Positioning

Positioning of tankers at the fill site (and unloading site) is of prime concern to the site officer. If necessary, tankers should be backed into the fill areas since they are generally much easier to maneuver when empty. The next-arriving tanker should be positioned so it does not block other units or filling positions. In addition, all tankers should be located within reach of the discharge lines from the fill pumper. Normally, tankers should be filled one at a time per fill site pumper. This will enable each tanker to load faster and possibly prevent a traffic jam at the unloading site. The second tanker should be hooked up as soon as possible so that it can begin filling as soon as the first one is full. There are times when two lines from the fill site pumper may go into one tanker and one or two additional lines into a second tanker. When the first unit is full, a line can be transferred to the second and third tanker positioned in the open space. The make and break personnel should also be instructed to open and close valves and hose clamps slowly to prevent water hammer (Figure 7.38). If there are delays in the shuttle, tankers can be dispatched to an open position from a staging area located as close to the fill site as possible. The fill site officer should call for tankers from the staging area as needed, with the highest fill and dump rate tankers given priority. The fill site officer or water supply officer

Turnouts can be used to improve tanker safety when a narrow road must be used for a tanker shuttle. Driveways or wide lanes can serve as turnouts anywhere along the shuttle route (Figure 7.40). Full tankers moving to the dump site should broadcast their position when approaching narrow road segments. Empty tankers near the location should yield the right-of-way by pulling into a turnout until the full tanker has passed.

Long driveways may complicate the response of water supply apparatus. They often require a pumper to draft from a reservoir or nurse tanker at the base of the drive and relay water to the attack pumper(s) (Figure 7.41). Portable tanks or nurse tankers should not block the main road.

Nonfire department designed tankers must be handled with a much higher degree of safety consciousness. Many of these commercial tankers are not baffled, many have a high center of gravity, and can be a severe hazard if driven recklessly. All tanker operators should be trained in proper driving skills for that size vehicle.

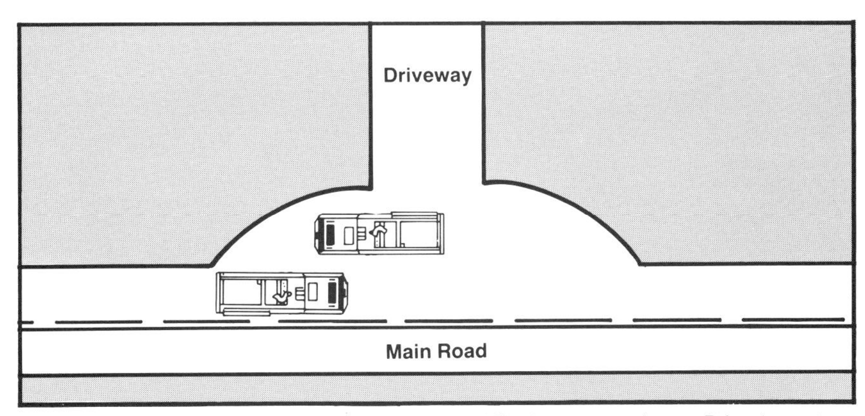

Figure 7.40 Passing other apparatus can be very difficult on narrow lanes. Driver/operators should take advantage of turnouts or other wide spots in the road for passing.

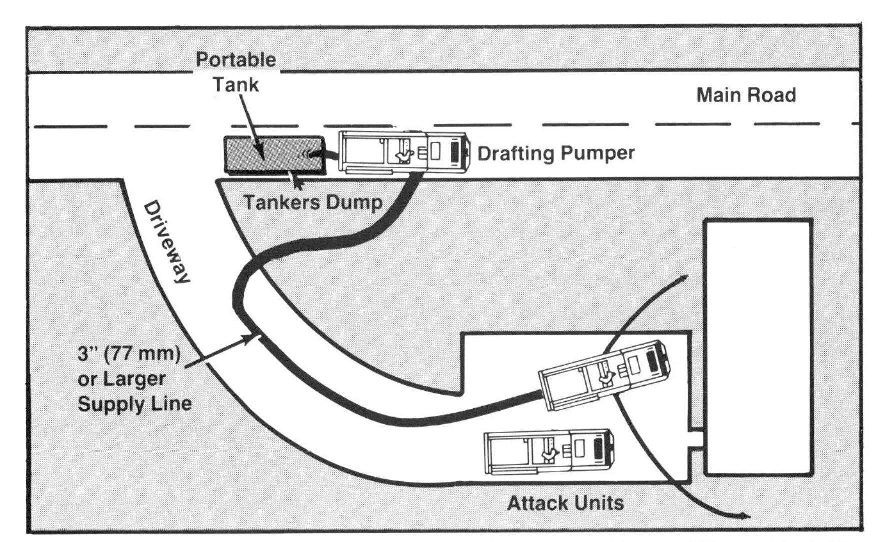

Figure 7.41 When a fire involves operating down a long narrow driveway, it is most desirable for attack apparatus to lay hoselines down the drive and for the tanker unloading site to be established out on the main road.

Unloading Methods

Water may be unloaded at a fire scene in four basic ways:

- Using pump on tank vehicle (Figure 7.42)
- Using pump on water supply or attack pumper (Figure 7.43)
- Using large diameter dump valve (Figure 7.44)
- Pumping and dumping simultaneously (Figure 7.45)

Tankers must be equipped with an adequate size pump and piping in order to pump their water to the attack pumpers. Tankers may pump to the main or auxiliary pump inlets, may fill the pumper through the top fill opening, or use a direct tank fill inlet. The disadvantage of this operation is that the tanker must remain connected. A loss of time is created when the attack pumper cannot immediately accept the tanker's capacity. Time loss may also occur when making and breaking hose connections.

Figure 7.42 Some tankers have the capability to pump their load off.

Figure 7.43 Pumpers may draft directly from large nurse tankers.

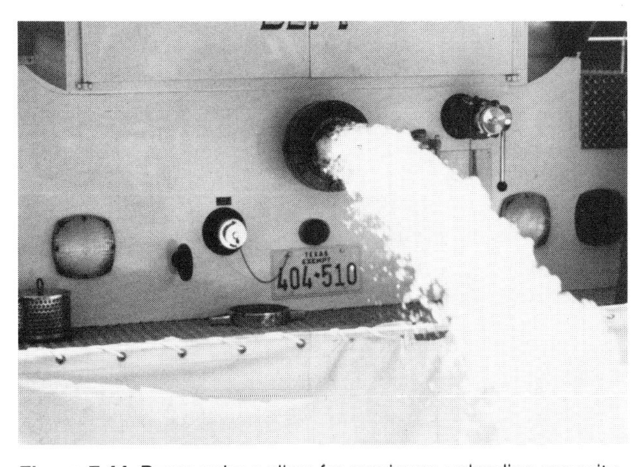

Figure 7.44 Dump valves allow for maximum unloading capacity. They may be located on any side of the apparatus but are most commonly found on the rear.

Figure 7.45 Pumper-tankers may be able to increase unloading time by simultaneously utilizing the dump valve and a pump discharge to remove water.

Tankers may pump their water into a portable tank (or nurse tanker) through one or a combination of hard suctions, large diameter hose, or standard hoselines. Longer hoselines may be needed depending on positioning. A higher volume pump is recommended for this type of transfer and hose connections must be made again unless pumping directly into the portable tank with stream shapers (Figure 7.46). The limiting factors are flow capacity of the tank to pump lines, pump size, venting capability, and the ability to control hoselines at the delivery point. Devices similar to those used for controlling line reaction when filling tankers should be used to free firefighters from having to hold the hoselines (Figures 7.47a-c on next page).

Another method of unloading tankers is for a pumper to draft water from the tanker. If fire streams are flowing, the attack

Courtesy of Matt Shields

Figure 7.46 Stream shapers are used to ensure that all water is directed toward the portable tank. They are most commonly used when water is being pumped off the apparatus.

Figure 7.47a A portable fill pipe such as this one can be used when hoselines are being used to fill portable tanks. *Courtesy of Marv Ackerman.*

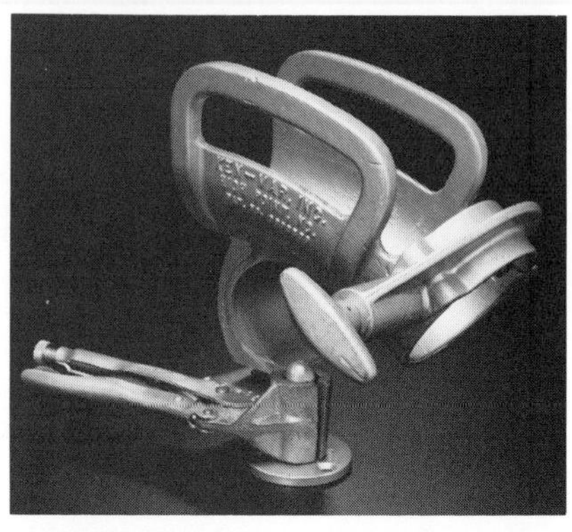

Figure 7.47b This commercially built clamp can be used to direct hoselines into a portable tank. *Courtesy of Ken-Mar, Inc.*

Figure 7.47c Homemade devices can be made to direct water into portable tanks.

pumper must use a gated intake or suction siamese to prevent interrupting the discharge lines. A supply pumper at a remote location may also be used to draft from the tanker and pump to the attack pumper. Tankers used as a draft source must have a threaded outlet(s) and sufficient tank venting to maintain the desired flow rate. Time will be lost when tankers must sit until their water is removed and suction hose disconnected.

The most efficient method of unloading tankers is to dump the water into portable tanks. Dumping, with sufficient portable tank capacity, will allow tankers to deliver their water rapidly and get back on the road. Both gravity and jet assisted dumps can be used to provide rapid flow rates (Figure 7.48).

Figure 7.48 Jet dumps increase the flow rate of any dump valve. *Courtesy of Tower Fire Apparatus.*

Gravity dumps employ large diameter valves and pressure created by the height of the water column in the tank to move the water. Discharge pressure will be based on column height.

1 foot (0.3 m) column = .434 psi (3 kPa)
2 foot (0.6 m) column = .868 psi (6 kPa)
3 foot (0.9 m) column = 1.302 psi (9 kPa)
4 foot (1.2 m) column = 1.736 psi (12 kPa)

The dumping time for an average tanker can be approximated by using 1 psi (6.9 kPa) as the average discharge pressure for the entire dump. Flows at this pressure for the various circular outlet sizes are given in Table 7.1.

TABLE 7.1
THEORETICAL FLOWS FROM CIRCULAR OUTLETS

Outlet		Flow
2½-inch (65 mm)	—	187 gpm (708 L/min)
4-inch (100 mm)	—	478 gpm (1 809 L/min)
6-inch (150 mm)	—	1,074 gpm (4 066 L/min)
8-inch (200 mm)	—	1,910 gpm (7 230 L/min)
10-inch (250 mm)	—	2,970 gpm (11 243 L/min)
12-inch (300 mm)	—	4,277 gpm (16 190 L/min)

It can be seen that large piping will geometrically increase the available water flow. The flow for a 10-inch (250 mm) square dump would be approximately 3,600 gpm (13 627 L/min). The theoretical flows are not realistic since there would not be an average of 1 psi for the entire load and on the bottom end the flow would decrease appreciably. Further, the flows indicated would be those achieved under ideal conditions. Most apparatus tanks are not designed with baffling and venting that allows these unloading rates. At the larger flows the 1 psi available is quickly used to overcome friction loss. Flows found in the field are listed in Table 7.2. Jet dumps employ the use of in-line jets, similar to those used in fill lines and siphons, for creating a venturi effect increasing water flow. These outlet jets are an integral part of a tanker's dump piping. An in-line jet can also be used to unload a tanker or auxiliary tank vehicle (Figures 7.49a-c).

TABLE 7.2
ACTUAL FLOWS FROM CIRCULAR OUTLETS

Outlet		Flow
4½-inch (115 mm)	—	450 gpm (1 743 L/min)
5-inch (125 mm)	—	550 gpm (2 0802L/min)
6-inch (150 mm)	—	650 gpm (2 461 L/min)
8-inch (200 mm)	—	1,200 gpm (4 542 L/min)
10-inch (250 mm)	—	1,600 gpm (6 057 L/min)
12-inch (300 mm)	—	2,300 gpm (8 706 L/min)
10-inch (square) (250 mm)	—	1,900 gpm (7 192 L/min)
16-inch (square) (400 mm)	—	4,400 gpm (16 656 L/min)

Figure 7.49a A commercially built in-line jet. *Courtesy of Ziamatic Corp.*

Figure 7.49b A homemade in-line jet.

Figure 7.49c In-line jets may be used to transfer water between portable tanks, or to speed the unloading of large tankers that are not equipped with a pump or jet dump valve. *Courtesy of Ziamatic Corp.*

An unloading operation that can save time is to simultaneously dump the tanker through a large valve and use its pump to discharge water through hoselines. Discharge lines must be left in place at the unloading site to reduce make and break time (Figure 7.50).

All dumps, whether jet assisted or not, should be installed to obtain the maximum pressure from the tank height. This can include sloping dump lines into the bottom of the tank, extending

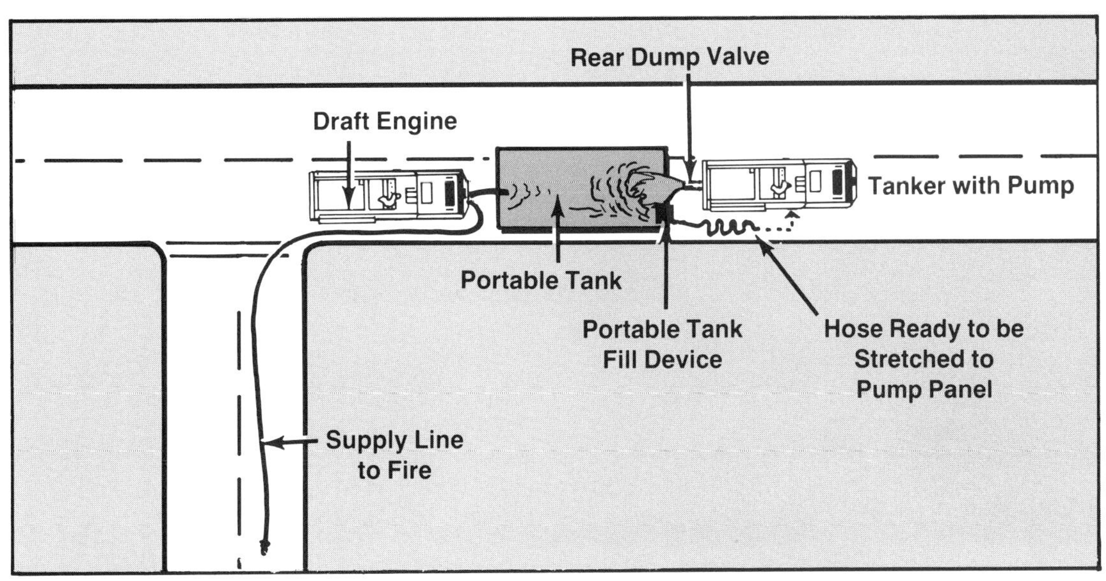

Figure 7.50 Laying out hose ahead of time makes operations at a dump and pump unloading site much smoother.

the dump from a sump below the tank, or having the dump valve built into the end of a trough that extends under the entire length of the tank.

Operations using gravity and jet dumps will provide the highest and most consistent flows to the fireground. This is due to the ability to return the tankers for more water quickly. Unloading site operations can involve multiple portable tanks and several dumping tankers in a variety of layouts and operational positions. Appendix D, Standard Operating Procedures, should be reviewed for possible unloading positions. Dumping operations can be further augmented through the use of supplemental devices. These may include air-operated solenoid valves that can operate dump valves from the tanker cab (Figure 7.51). Extended and movable dump chutes, piping, or hose will give extended rear

Figure 7.51 The air-operated solenoid valve in the cab of this tanker allows the driver/operator to dump the water without having to leave the cab.

dump reach and allow discharge off either side of the tanker (Figure 7.52). Tankers have even been manufactured with tank hoists that provide more rapid and complete dumping (Figure 7.53).

Figure 7.52 A variety of methods may be employed to channel water in the desired direction. *Courtesy of Marv Ackerman.*

Figure 7.53 Some tankers may be equipped with hydraulic hoists that elevate the tank during dumping operations to speed the unloading time. *Courtesy of Monroe Truck Equipment, Inc.*

Like loading operations, unloading operations require personnel to make and break connections, control valves, operate pumps, and open vents. Baffles must have sufficiently sized openings to allow free water movement at the bottom of the tank and air movement at the top during rapid filling or unloading (Figure 7.54).

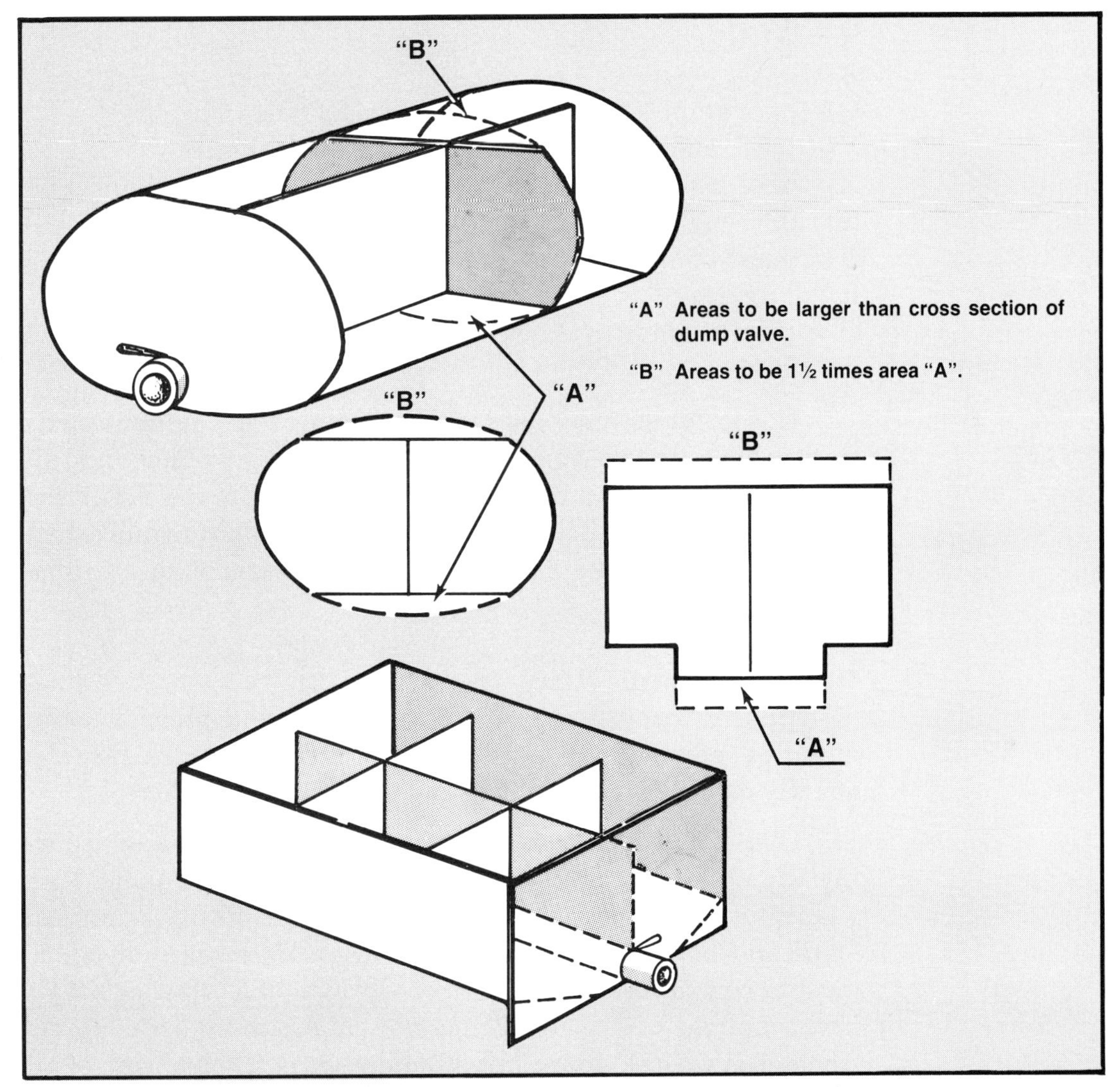

Figure 7.54 Proper baffling of tanks restricts water movement during transit, but allows for maximum loading and unloading capabilities.

Many small- and medium-sized residential fires can be controlled and extinguished with one pumper and one or two tankers. The fireground commander must avoid committing a tanker to a direct hookup since a portable tank can be set up in as few seconds, serves the same function, provides a dump point, and water reserve, if one tanker proves insufficient. The fireground commander must be ready to appoint a water supply officer to call on and manage all available water resources. Large fires or major incidents require a water supply officer and support network to effectively control multicompany water supply operations.

MAXIMUM CONTINUOUS FLOW CAPABILITY

The water supply officer can determine the maximum fire flow available from a particular shuttle setup. This information will be very valuable to the fireground commander as well as aiding the water supply officer in deciding when additional apparatus will be needed. The formula is

$$Q = \frac{V}{A+B+C}$$

Q = maximum continuous flow in gpm (L/min)
V = total tanker delivery capacity in gallons (liters)
A = time in minutes to empty tanker (including hose connection and handling time)
B = time in minutes to fill tanker (including hose connection and handling time)
C = time in minutes of average round trip travel from source to fire source

This formula can be used in determining the maximum shuttle flow by taking the sum of all operating tanker volumes and dividing it by the time it takes to make one complete cycle of all the operating tankers. This formula can also be applied to individual tankers to determine their efficiency for various shuttle distances.

MUTUAL AID SHUTTLES

Fire departments may often need to call on neighboring communities or districts for assistance. Water supply shuttles are frequently run with several departments working together.

There are several considerations that must be evaluated during drills before the need arises to call for mutal aid tankers. Varying types of equipment between departments must be examined for fill and dump site considerations. Hose and tanker connections must be compatible or adapters readily available. Quick-connect hose couplings will reduce connecting time, but require adapters when used with standard thread connections. The location of fill openings on mutual aid apparatus will vary and may require a different type of fill setup. The arrangement for unloading may also be different and require different positioning of portable tanks or nurse tankers at the unloading site.

The capability of mutual aid tankers will determine the number of trucks required and how they will be used in the shuttle operation (shuttling, as a nurse, or in reserve). These capabilities should be learned in drills with the information readily available to the water supply officer when a fire occurs (Figure 7.55).

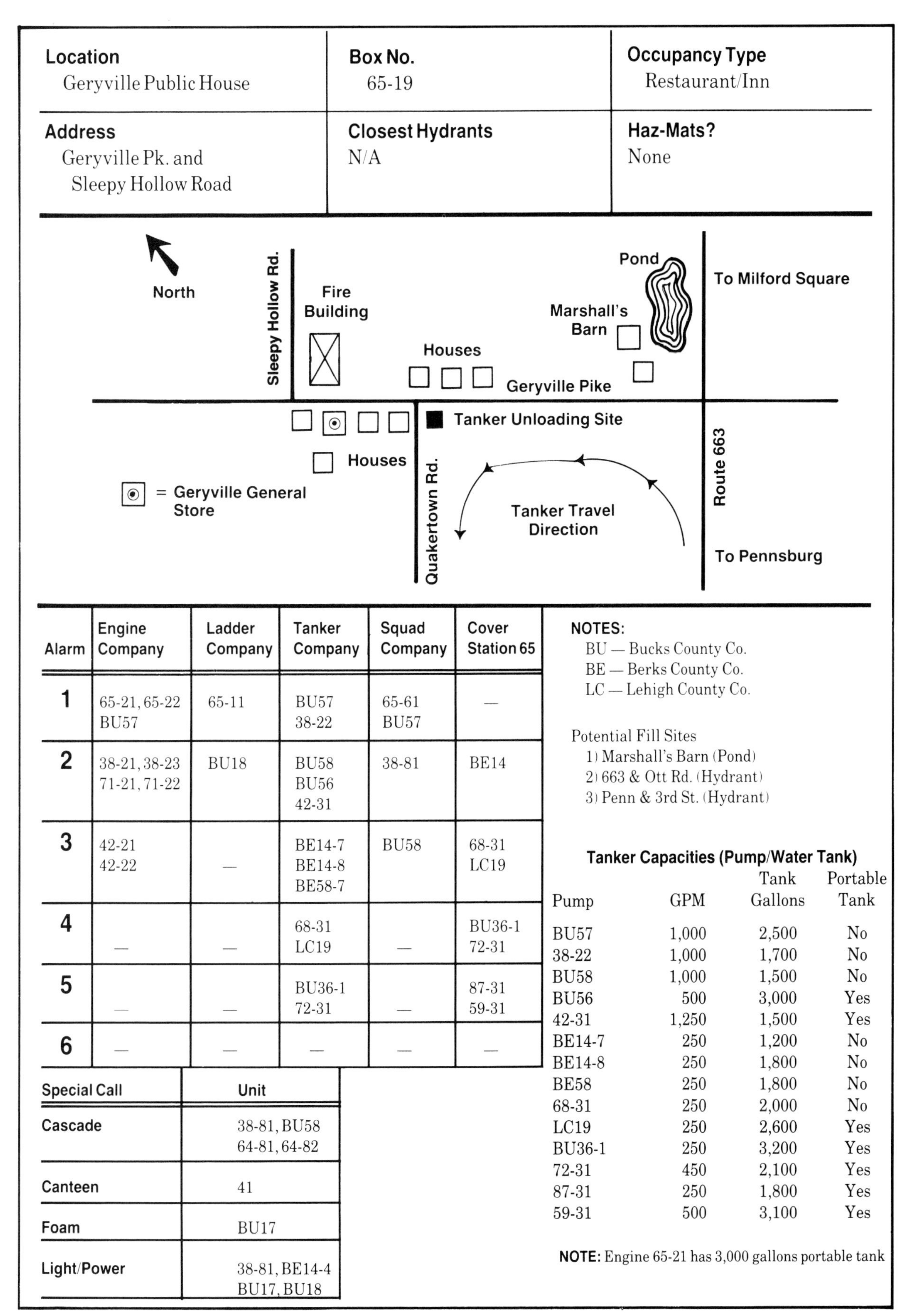

Location	Box No.	Occupancy Type
Geryville Public House	65-19	Restaurant/Inn
Address	**Closest Hydrants**	**Haz-Mats?**
Geryville Pk. and Sleepy Hollow Road	N/A	None

Alarm	Engine Company	Ladder Company	Tanker Company	Squad Company	Cover Station 65
1	65-21, 65-22 BU57	65-11	BU57 38-22	65-61 BU57	—
2	38-21, 38-23 71-21, 71-22	BU18	BU58 BU56 42-31	38-81	BE14
3	42-21 42-22	—	BE14-7 BE14-8 BE58-7	BU58	68-31 LC19
4	—	—	68-31 LC19	—	BU36-1 72-31
5	—	—	BU36-1 72-31	—	87-31 59-31
6	—	—	—	—	—

Special Call	Unit
Cascade	38-81, BU58 64-81, 64-82
Canteen	41
Foam	BU17
Light/Power	38-81, BE14-4 BU17, BU18

NOTES:
BU — Bucks County Co.
BE — Berks County Co.
LC — Lehigh County Co.

Potential Fill Sites
1) Marshall's Barn (Pond)
2) 663 & Ott Rd. (Hydrant)
3) Penn & 3rd St. (Hydrant)

Tanker Capacities (Pump/Water Tank)

Pump	GPM	Tank Gallons	Portable Tank
BU57	1,000	2,500	No
38-22	1,000	1,700	No
BU58	1,000	1,500	No
BU56	500	3,000	Yes
42-31	1,250	1,500	Yes
BE14-7	250	1,200	No
BE14-8	250	1,800	No
BE58	250	1,800	No
68-31	250	2,000	No
LC19	250	2,600	Yes
BU36-1	250	3,200	Yes
72-31	450	2,100	Yes
87-31	250	1,800	Yes
59-31	500	3,100	Yes

NOTE: Engine 65-21 has 3,000 gallons portable tank

Figure 7.55 (U.S.) Box assignments should contain information on tanker capabilities so adequate aid can be sent when requested.

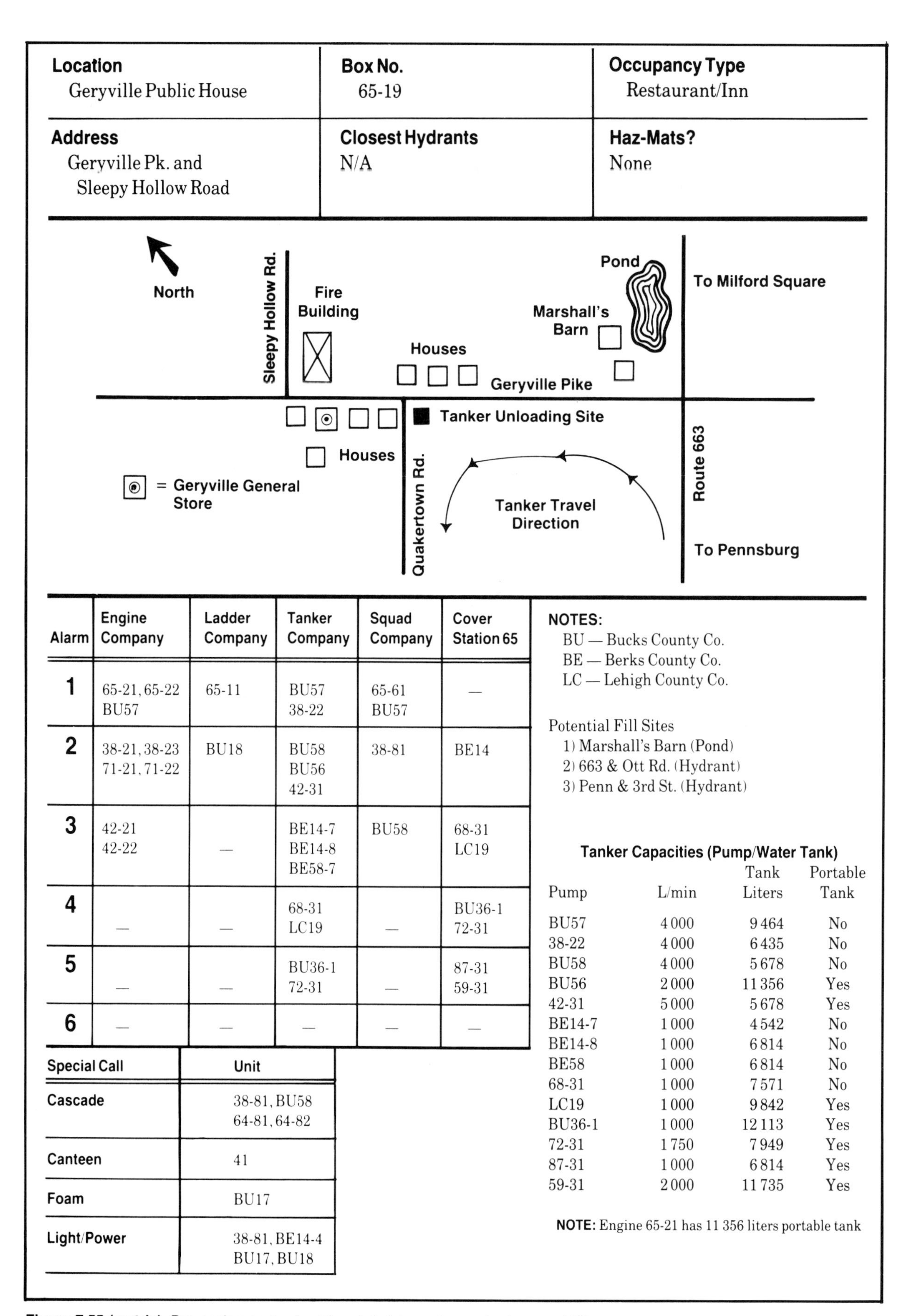

Location	Box No.	Occupancy Type
Geryville Public House	65-19	Restaurant/Inn
Address	**Closest Hydrants**	**Haz-Mats?**
Geryville Pk. and Sleepy Hollow Road	N/A	None

Alarm	Engine Company	Ladder Company	Tanker Company	Squad Company	Cover Station 65
1	65-21, 65-22 BU57	65-11	BU57 38-22	65-61 BU57	—
2	38-21, 38-23 71-21, 71-22	BU18	BU58 BU56 42-31	38-81	BE14
3	42-21 42-22	—	BE14-7 BE14-8 BE58-7	BU58	68-31 LC19
4	—	—	68-31 LC19	—	BU36-1 72-31
5	—	—	BU36-1 72-31	—	87-31 59-31
6	—	—	—	—	—

Special Call	Unit
Cascade	38-81, BU58 64-81, 64-82
Canteen	41
Foam	BU17
Light/Power	38-81, BE14-4 BU17, BU18

NOTES:
BU — Bucks County Co.
BE — Berks County Co.
LC — Lehigh County Co.

Potential Fill Sites
1) Marshall's Barn (Pond)
2) 663 & Ott Rd. (Hydrant)
3) Penn & 3rd St. (Hydrant)

Tanker Capacities (Pump/Water Tank)

Pump	L/min	Tank Liters	Portable Tank
BU57	4 000	9 464	No
38-22	4 000	6 435	No
BU58	4 000	5 678	No
BU56	2 000	11 356	Yes
42-31	5 000	5 678	Yes
BE14-7	1 000	4 542	No
BE14-8	1 000	6 814	No
BE58	1 000	6 814	No
68-31	1 000	7 571	No
LC19	1 000	9 842	Yes
BU36-1	1 000	12 113	Yes
72-31	1 750	7 949	Yes
87-31	1 000	6 814	Yes
59-31	2 000	11 735	Yes

NOTE: Engine 65-21 has 11 356 liters portable tank

Figure 7.55 (metric) Box assignments should contain information on tanker capabilities so adequate aid can be sent when requested.

Chapter 7 Review

Answers on page 262

Answers on page 262

TRUE-FALSE. Mark each statement true or false. If false, explain why.

1. Water has a 20 percent greater weight by volume than fuel oil.

   ☐ T ☐ F ______________________________

   ____

2. Standard pumps on fuel oil tankers serve very well for fire fighting purposes.

   ☐ T ☐ F ______________________________

   ____

3. The main disadvantage of the four-corner post type portable tank is that it is too large to fit into one lane of a road.

   ☐ T ☐ F ______________________________

   ____

4. When readying a pumper to draft from a supply source, the hard suction hose should be connected to the apparatus prior to moving to the final drafting position.

   ☐ T ☐ F ______________________________

   ____

5. Top filling is the most desirable fill method for tankers.

   ☐ T ☐ F ______________________________

   ____

6. Tankers with the highest fill rates should be given priority at the fill site because they require less time to fill and unload.

   ☐ T ☐ F ______________________________

   ____

MULTIPLE CHOICE: Circle the correct answer.

7. Apparatus with a tank capacity of greater than ________ gallons (liters) should use either tandem rear axles or semi-trailer construction.
   A. 1,000 gallons (3 785 L)
   B. 1,250 gallons (4 730 L)
   C. 1,500 gallons (5 678 L)
   D. 2,000 gallons (7 570 L)

8. Tankers carrying greater than ________ gallons (liters) are best utilized as nurse-tankers.
   A. 2,500 gallons (9 464 L)
   B. 3,000 gallons (11 360 L)
   C. 4,000 gallons (15 142 L)
   D. 5,000 gallons (18 930 L)

9. Which one of the following factors will *not* have an effect on the roadability of a good fire department tanker?
   A. Proper tank baffling
   B. Adequate warning devices
   C. Adequate power
   D. Adequate chassis

10. Which of the following methods is *not* recommended for transferring water between two portable tanks?
    A. Tank connectors
    B. Jet siphons
    C. Siphons
    D. Four firefighters using large buckets

11. Which of the following factors will *not* decrease the fill rate of a tanker?
    A. Altitude
    B. Inadequate tank baffling
    C. Improper baffling
    D. Inadequate source

12. Which is generally the quickest unloading method?
    A. Dump it off
    B. Pump it off
    C. Draft it off
    D. All are about the same

LISTING

13. List three ways pumps can be used in water shuttle operations.

____

____

____

14. List two problems associated with large tankers.

____

____

15. List two factors that must be considered when deciding whether it is feasible to convert another type of tank truck to a fire department tanker.

____

____

16. List three construction requirements needed to make a piece of apparatus a safe and efficient water mover.

____

____

____

17. List three types of fill methods for tankers.

____

____

____

FILL IN THE BLANK: Fill in the blanks with the correct response.

18. A pumper should draft from a portable tank from the corner of the tank or ____________ .

19. The three basic methods of unloading water at the fire scene are to pump it off, dump it off, and ________ it off.

SHORT ANSWER: Answer each item briefly.

20. Describe the main reason a single portable pump may not be suitable for filling tankers in a water shuttle operation.

____

____

21. Name two types of portable tanks.

____

____

22. Folding style portable tanks may be stored on the apparatus in a number of ways. Name two.

____

____

23. In addition to tanker shuttle operations, name two other situations in which portable tanks may be used.

____

____

24. Name at least three areas of accident potential in the route of a tanker shuttle.

____

____

____

25. There are two types of tanker dump devices. Name them and tell which is quickest.

____

____

Appendix A
Excerpts from *Uniform Fire Code*

APPENDIX II
FIRE-FLOW REQUIREMENTS FOR BUILDINGS

1. **SCOPE**

This appendix is the procedure for determining fire-flow requirements for all buildings or portions of buildings hereafter constructed. This appendix is not intended to apply to structures other than buildings. The fire-flow requirement is the quantity of water in gallons per minute needed to control an anticipated fire in a building or group of buildings. The chief shall establish the minimum residual pressure and the flow duration to be used when determining fire flow.

2. **DEFINITIONS**

FIRE AREA is the total floor area in square feet for all floor levels within the exterior walls, or under the horizontal projection of the roof of a building. Four-hour area separation walls may be considered as dividing a building into separate fire areas for the purpose of determining fire flow.

3. **MODIFICATIONS**

Fire-flow requirements may be modified downward for isolated buildings in rural areas or small communities where the development of full fire-flow requirements is impractical.

Fire flow may be modified upward where conditions indicate an unusual susceptibility to group fires or conflagrations. An upward modification shall not be more than twice that required for the building under construction.

4. **FIRE-FLOW REQUIREMENTS FOR BUILDINGS**

The minimum fire-flow requirements for one- and two-family dwellings shall be 1000 gallons per minute.

> **EXCEPTION:** Fire flow may be reduced 50 percent when the building is provided with an approved automatic sprinkler system.

The fire flow for buildings other than one- and two-family dwellings shall be not less than that specified in Table No. 4-A.

> **EXCEPTION:** The required fire flow may be reduced up to 75 percent when the building is provided with an approved automatic sprinkler system, but in no case less than 1500 gallons per minute.

In Types I and II-F.R. construction, only the three largest successive floor areas shall be used.

TABLE NO. 4-A
FIRE-FLOW GUIDE FOR BUILDINGS OTHER THAN ONE- AND TWO-FAMILY DWELLINGS

Fire Flow (Gallons Per Minute)	CONSTRUCTION TYPE				
	I II-F.R.	II-One-HR. III-ONE-HR.	IV-H.T. V-ONE-HR.	II-N III-N	V-N
	TOTAL FIRE AREA IN SQUARE FEET				
1,500	22,700	12,700	8,200	5,900	3,600
1,750	30,200	17,000	10,900	7,900	4,800
2,000	38,700	21,800	12,900	9,800	6,200
2,250	48,300	24,200	17,400	12,600	7,700
2,500	59,000	33,200	21,300	15,400	9,400
2,750	70,900	39,700	25,500	18,400	11,300
3,000	83,700	47,100	30,100	21,800	13,400
3,250	97,700	54,900	35,200	25,900	15,600
3,500	112,700	63,400	40,600	29,300	18,000
3,750	128,700	72,400	46,400	33,500	20,600
4,000	145,900	82,100	52,500	37,900	23,300
4,250	164,200	92,400	59,100	42,700	26,300
4,500	183,400	103,100	66,000	47,700	29,300
4,750	203,700	114,600	73,300	53,000	32,600
5,000	225,200	126,700	81,100	58,600	36,000
5,250	247,700	139,400	89,200	65,400	39,600
5,500	271,200	152,600	97,700	70,600	43,400
5,750	295,900	166,500	106,500	77,000	47,400
6,000	UNLIMITED	UNLIMITED	115,800	83,700	51,500
6,250	"	"	125,500	90,600	55,700
6,500	"	"	135,500	97,900	60,200
6,750	"	"	145,800	106,800	64,800
7,000	"	"	156,700	113,200	69,600
7,250	"	"	167,900	121,300	74,600
7,500	"	"	179,400	129,600	79,800
7,750	"	"	191,400	138,300	85,100
8,000	"	"	UNLIMITED	UNLIMITED	UNLIMITED

STAFF NOTE: Fire-Flow Appendix will be Appendix II. All other following appendices will be renumbered.

MODEL CODE CONVERSION CHART

The following chart enables users of other building and life safety codes to utilize the *Uniform Fire Code* material, which is based on the *Uniform Fire Code* building construction types.

U.B.C. Construction Type	OTHER MODEL CODE CONSTRUCTION TYPE		
	Basic/National Building Code (BOCA)	Standard Building Code (SBCCI)	NFPA 220 (NFC)
I-F.R.	1A	I	I (443)
II-F.R.	1B, 2A	II	I (332), II (222)
II-1 hr.	2B	IV-1hr.	II (111)
II-N	2C	IV-Unprotected	II (000)
III-1 hr.	3A	V-1 hr.	III (211)
III-N	3B	V-Unprotected	III (200)
IV-H.T.	4	III	IV (2 HH)
V-1 hr.	5A	VI-1 hr.	V (111)
V-N	5B	VI-Unprotected	V (000)

ISO *Guide for Determination of Required Fire Flow*

This guide has been prepared for the use of the municipal survey and grading personnel of Insurance Services Office and other fire insurance rating organizations. It is being made available to municipal officials, consulting engineers, and other interested parties as an aid in estimating fire flow requirements. It should be recognized that this publication is a "guide" in the true sense of the word, and requires a certain amount of knowledge and experience in fire protection engineering for its effective application.

INSURANCE SERVICES OFFICE
160 Water Street
New York, New York 10038
December 1974

ISO COMMERCIAL RISK SERVICES, INC.
2 Sylvan Way
Parsippany, New Jersey 07054

INSURANCE SERVICES OFFICE

Guide For Determination of Required Fire Flow

1. An estimate of the fire flow required for a given fire area may be determined by the formula:

$$F = 18\,C\,(A)^{0.5}$$

where

F = the required fire flow in gpm

C = coefficient related to the type of construction

C = 1.5 for wood frame construction
= 1.0 for ordinary construction
= 0.9 for heavy timber type buildings
= 0.8 for noncombustible construction
= 0.6 for fire-resistive construction

Note: For types of construction and/or materials that do not fall within the categories given, use a coefficient reflecting the difference. Coefficients shall not be greater than 1.5 nor less than 0.6 and may be determined by interpolation. Such interpolation shall be between consecutive types of construction as listed above. Definitions of types of construction are included in this Appendix.

A = The total floor area (including all stories, but excluding basements) in the building being considered. For fire-resistive buildings consider the 6 largest successive floor areas if the vertical openings are unprotected; if the vertical openings are properly protected, consider only the 3 largest successive floor areas.

The fire flow as determined by the above shall not exceed:

8,000 gpm for wood frame construction

8,000 gpm for ordinary and heavy timber construction
6,000 gpm for noncombustible construction
6,000 gpm for fire-resistive construction

except that for a normal 1-story building of any type of construction the fire flow shall not exceed 6,000 gpm.

The fire flow shall not be less than 500 gpm.

For 1-family and small 2-family dwellings not exceeding 2 stories in height see note 10.

2. The value obtained in No. 1 above may be reduced by up to 25% for occupancies having a low fire hazard or may be increased by up to 25% for occupancies having a high fire hazard. As a guide for determining low or high hazard occupancies see the lists in Appendix C.

   The fire flow shall not be less than 500 gpm.

3. The value obtained in No. 2 above may be reduced by up to 50% for complete automatic sprinkler protection. Where buildings are either fire-resisitive or noncombustible construction, and have a low fire hazard, the reduction may be up to 75%. The percentage reduction made for an automatic sprinkler system will depend upon the extent to which the system is judged to reduce the possibility of fires spreading within and beyond the fire area. Normally this reduction will not be the maximum allowed without proper system supervision including water flow and valves.

4. To the value obtained in No. 2 above a percentage should be added for structures exposed within 150 feet by the fire area under consideration. This percentage shall depend upon the height, area, and construction of the building(s) being exposed, the separation, openings in the exposed building(s), the length of exposure, the provision of automatic sprinklers and/or outside sprinklers in the building(s) exposed, the occupancy of the exposed building(s), and the effect of hillside location on the possible spread of fire.

The percentage for any one side generally should not exceed the following limits for the separations shown:

Separation	Percentage
0 - 10 feet	25%
11 - 30	20
31 - 60	15
61 - 100	10
101 - 150	5

The total percentage shall be the sum of the percentages for all sides, but shall not exceed 75%.

5. The value obtained in No. 2 above is reduced by the percentage (if any) determined in No. 3 above and increased by the percentage (if any) determined in No. 4 above.

The fire flow shall not exceed 12,000 gpm nor be less than 500 gpm.

Note 1: The guide is not expected to necessarily provide an adequate value for lumber yards, petroleum storage, refineries, grain elevators, and large chemical plants but may indicate a minimum value for these hazards.

Note 2: Judgment must be used for business, industrial, and other occupancies not specifically mentioned.

Note 3: Consideration should be given to the configuration of the building(s) being considered and to the fire department accessibility.

Note 4: Wood frame structures separated by less than 10 feet shall be considered as one fire area.

Note 5: Party Walls – Normally an unpierced party (common) wall may warrant up to a 10% exposure charge.

Note 6: High one-story buildings – When a building is stated as 1 = 2, or more stories, the number of stories to be used in the formula depends upon the use being made of the building. For example, consider 1 = 3-story building. If the building is being used for high-piled stock, or for rack storage, the building would probably be considered as 3 stories and, in addition, an increased percentage for occupancy may be warranted. However, if the building is being used for steel fabrication and the extra height is provided only to facilitate movement of objects by a crane, the building would probably be considered as a 1-story building and a decreased percentage for occupancy may be warranted.

Note 7: If a building is exposed within 150 feet, normally some percentage increase for exposure will be made.

Note 8: Where wood shingle roofs could contribute to spreading fires, add 500 gpm.

Note 9: Any noncombustible building is considered to warrant an 0.8 coefficient.

Note 10: Dwellings – For groupings of 1-family and small 2-family dwellings not exceeding 2 stories in height, the following short method may be used. (For other residential buildings, the regular method should be used.)

Exposure distances	Suggested required fire flow
Over 100'	500 gpm
31 - 100'	750 - 1000
11 - 30'	1000 - 1500
10' or less	1500 - 2000*

*If the buildings are continuous, use a minimum of 2500 gpm. Also consider Note 8.

Outline of Procedure

A. Determine the type of construction.

B. Determine the ground floor area.

C. Determine the height in stories.

D. Using tables in this Appendix, determine the required fire flow to the nearest 250 gpm.

E. Determine the increase or decrease for occupancy and apply to the value obtained in D above. Do not round off the answer.

F. Determine the decrease, if any, for automatic sprinkler protection. Do not round off the value.

G. Determine the total increase for exposures. Do not round off the value.

H. To the answer obtained in E, subtract the value obtained in F and add the value obtained in G.

Round off the final answer to the nearest 250 gpm if less than 2500 gpm and to the nearest 500 gpm if greater than 2500 gpm.

Use of Tables (Steps A, B, C, D)

The tables use the GROUND AREA of the building and the height of the building in stories. Using the table corresponding to the type of construction, look under the number of stories and locate the ground area of the building(s) being considered between two ground areas given in the table. The corresponding fire flow is found in the left column.

Examples

a. Given: A 3-story building of ordinary construction of 7300 square feet (ground area). Using the table C = 1.0, in the 3-story column, 7300 square feet falls between 7100 and 8500 square feet and the corresponding fire flow is 2750 gpm.

b. Given: A 3-story building of ordinary construction of 7300 square feet (ground area) communicating to a 5-story building of ordinary construction of 9700 square feet (ground area) for a total ground area of 17,000 square feet. Determine the total floor area which equals 3 (7300) + 5 (9700) = 70,400 square feet. Using the table C = 1.0, under the one story column for 70,400 sq. ft. the corresponding fire flow is 4750 gpm.

c. Given: A 3-story wood frame building of 7300 square feet (ground area) communicating with a 5-story building of ordinary construction of 9700 square feet (ground area) for a total ground area of 17,000 square feet.

Determine the total floor area for each type of construction and for the fire area which is 3 (7300) = 21,900 square feet of wood frame construction, 5 (9700) = 48,500 square feet of ordinary construction, and a total area of 70,400 square feet with 31% being of wood frame construction and 69% being of ordinary construction. Under the one-story column in the wood frame construction table (C = 1.5), an area of 70,400 square feet has a corresponding fire flow of 7250 gpm. Similarly, under the one-story column in the ordinary construction table (C = 1.0), an area of 70,400 square feet has a corresponding fire flow of 4750 gpm. In this case, the fire flow will be 31% (7250) + 69% (4750) = 2250 + 3280 = 5530 gpm, or to the nearest 250 gpm, = 5500 gpm.

d. Given: A 2-story building consisting of 10,000 square feet (ground area) of wood frame construction, 15,000 square feet (ground area) of ordinary construction, 20,000 square feet (ground area) of noncombustible construction, and 25,000 square feet (ground area) of fire-resistive construction. The total floor area is 140,000 square feet. The maximum fire flow for wood frame construction is at 85,100 square feet (see table). **Note:** "When the total area exceeds the upper limit for the poorest type of construction, limit the floor area of the best type(s) of construction so that the total area considered does not exceed the upper limit for the poorest type of construction." Consider 2 x 10,000 = 20,000 square feet of wood frame construction, plus 2 x 15,000 = 30,000 square feet of ordinary construction plus 2 x 20,000 = 40,000 square feet (limited to 35,100 square feet) of noncombustible construction. The fire flow will be 24% x 8000 (wood frame) + 35% x 6750* (ordinary) + 41% x 5500* (noncombustible) = 1920 + 2362 + 2255 = 6537 gpm, or to the nearest 250 gpm, = 6500 gpm.

*based upon 140,000 square feet.

e. Given: A 2-story building of ordinary construction of 105,000 square feet (ground area) communicates with a 1-story building of noncombustible construction of 80,000 square feet (ground area). Normally the required fire flow would be determined by proportioning as in "c" above. This would result in a required fire flow of 7460 gpm, or 7500 gpm. However, it is to be noted that the total area of the 2-story building alone results in a fire flow of 8,000 gpm and, of course, the logical answer would be 8,000 gpm. Any time the total area results in the use of an upper limit for fire flow, the possibility of a portion of the fire area justifying the upper limit must be investigated.

f. Given: A normal 1-story building of ordinary construction of 210,000 square feet (ground area). The table gives a required fire flow of 8,000 gpm; however, since this is a normal 1-story building, the maximum fire flow is 6,000 gpm.

g. Given: A normal 1-story building of ordinary construction of 80,000 square feet communicates with a normal 1-story building of noncombustible construction of 85,000 square feet. Normally the required fire flow would be determined by proportioning as in "c" above. This would result in a required fire flow of 6480 gpm, or 6500 gpm. However, since these are normal 1-story buildings the maximum fire flow is 6,000 gpm.

REQUIRED FIRE FLOW TABLES

INSURANCE SERVICES OFFICE
FIRE FLOW vs. GROUND AREA

Wood Frame Construction
(Ground area in square feet)

$F = 18C(A)^{0.5}$

F = gpm; $C = 1.5$

A = area in sq. ft.

gpm	1	2	3	4	5	6 Stories
500						
	500	300	200	100	100	100
750						
	1,100	600	400	300	200	200
1000						
	1,700	900	600	400	300	300
1250						
	2,600	1,300	900	700	500	400
1500						
	3,600	1,800	1,200	900	700	600
1750						
	4,800	2,400	1,600	1,200	1,000	800
2000						
	6,200	3,100	2,100	1,600	1,200	1,000
2250						
	7,700	3,900	2,600	1,900	1,500	1,300
2500						
	9,400	4,700	3,100	2,400	1,900	1,600
2750						
	11,300	5,700	3,800	2,800	2,300	1,900
3000						
	13,400	6,700	4,500	3,400	2,700	2,200
3250						
	15,600	7,800	5,200	3,900	3,100	2,600
3500						
	18,000	9,000	6,000	4,500	3,600	3,000
3750						
	20,600	10,300	6,900	5,200	4,100	3,400
4000						
	23,300	11,700	7,800	5,800	4,700	3,900
4250						

gpm	1	2	3	4	5	6 Stories
4250						
	26,300	13,200	8,800	6,600	5,300	4,400
4500						
	29,300	14,700	9,800	7,300	5,900	4,900
4750						
	32,600	16,300	10,900	8,200	6,500	5,400
5000						
	36,000	18,000	12,000	9,000	7,200	6,000
5250						
	39,600	19,800	13,200	9,900	7,900	6,600
5500						
	43,400	21,700	14,500	10,900	8,700	7,200
5750						
	47,400	23,700	15,800	11,900	9,500	7,900
6000						
	51,500	25,800	17,200	12,900	10,300	8,600
6250						
	55,700	27,900	18,600	13,900	11,100	9,300
6500						
	60,200	30,100	20,100	15,100	12,000	10,000
6750						
	64,800	32,400	21,600	16,200	13,000	10,800
7000						
	69,600	34,800	23,200	17,400	13,900	11,600
7250						
	74,600	37,300	24,900	18,700	14,900	12,400
7500						
	79,800	39,900	26,600	20,000	16,000	13,300
7750						
	85,100	42,600	28,400	21,300	17,000	14,200
8000						

INSURANCE SERVICES OFFICE
FIRE FLOW vs. GROUND AREA

Ordinary Construction
(Ground area in square feet)

$F = 18C(A)^{0.5}$
F = gpm; C = 1.0
A = area in sq. ft.

gpm	1	2	3	4	5	6 Stories
500						
	1,200	600	400	300	200	200
750						
	2,400	1,200	800	600	500	400
1000						
	3,900	2,000	1,300	1,000	800	700
1250						
	5,800	2,900	1,900	1,500	1,200	1,000
1500						
	8,200	4,100	2,700	2,100	1,600	1,400
1750						
	10,900	5,500	3,600	2,700	2,200	1,800
2000						
	13,900	7,000	4,600	3,500	2,800	2,300
2250						
	17,400	8,700	5,800	4,400	3,500	2,900
2500						
	21,300	10,700	7,100	5,300	4,300	3,600
2750						
	25,500	12,800	8,500	6,400	5,100	4,300
3000						
	30,100	15,100	10,000	7,500	6,000	5,000
3250						
	35,200	17,600	11,700	8,800	7,000	5,900
3500						
	40,600	20,300	13,500	10,200	8,100	6,800
3750						
	46,400	23,200	15,500	11,600	9,300	7,700
4000						
	52,500	26,300	17,500	13,100	10,500	8,800
4250						
	59,100	29,600	19,700	14,800	11,800	9,900
4500						

gpm	1	2	3	4	5	6 Stories
4500						
	66,000	33,000	22,000	16,500	13,200	11,000
4750						
	73,300	36,700	24,400	18,300	14,700	12,200
5000						
	81,100	40,600	27,000	20,300	16,200	13,500
5250						
	89,200	44,600	29,700	22,300	17,800	14,900
5500						
	97,700	48,900	32,600	24,400	19,500	16,300
5750						
	106,500	53,300	35,500	26,600	21,300	17,800
6000						
	115,800	57,900	38,600	28,900	23,200	19,300
6250						
	125,500	62,800	41,800	31,400	25,100	20,900
6500						
	135,500	67,800	45,200	33,900	27,100	22,600
6750						
	145,800	72,900	48,600	36,500	29,200	24,300
7000						
	156,700	78,400	52,200	39,200	31,300	26,100
7250						
	167,900	84,000	56,000	42,000	33,600	28,000
7500						
	179,400	89,700	59,800	44,900	35,900	29,900
7750						
	191,400	95,700	63,800	47,900	38,300	31,900
8000						

INSURANCE SERVICES OFFICE
FIRE FLOW vs. GROUND AREA

$F = 18C(A)^{0.5}$

F = gpm; $C = 0.8$

A = area in sq. ft.

Noncombustible Construction (Ground area in square feet)

gpm	1	2	3	4	5	6 Stories
500						
	1,900	1,000	600	500	400	300
750						
	3,700	1,900	1,200	900	700	600
1000						
	6,100	3,100	2,000	1,500	1,200	1,000
1250						
	9,100	4,600	3,000	2,300	1,800	1,500
1500						
	12,700	6,400	4,200	3,200	2,500	2,100
1750						
	17,000	8,500	5,700	4,100	3,400	2,800
2000						
	21,800	10,900	7,300	5,500	4,400	3,600
2250						
	27,200	13,600	9,100	6,800	5,400	4,500
2500						
	33,200	16,600	11,100	8,300	6,600	5,500
2750						
	39,700	19,900	13,200	9,900	7,900	6,600
3000						
	47,100	23,600	15,700	11,800	9,400	7,900
3250						
	54,900	27,500	18,300	13,700	11,000	9,200
3500						
	63,400	31,700	21,100	15,900	12,700	10,600
3750						
	72,400	36,200	24,100	18,100	14,500	12,100
4000						
	82,100	41,200	27,400	20,500	16,400	13,700
4250						
	92,400	46,200	30,800	23,100	18,500	15,400
4500						
	103,100	51,600	34,400	25,800	20,600	17,200
4750						
	114,600	57,300	38,200	28,700	22,900	19,100
5000						
	126,700	63,400	42,200	31,700	25,300	21,100
5250						
	139,400	69,700	46,500	34,900	27,900	23,200
5500						
	152,600	76,300	50,900	38,200	30,500	25,400
5750						
	166,500	83,300	55,500	41,600	33,300	27,800
6000						

INSURANCE SERVICES OFFICE
FIRE FLOW vs. GROUND AREA
Fire-Resistive Construction
(Ground area in square feet)

$F = 18C(A)^{0.5}$
F = gpm; C = 0.6
A = area in sq. ft.

gpm	1	2	3	4	5	6 Stories
500						
	3,300	1,700	1,100	800	700	600
750						
	6,600	3,300	2,200	1,700	1,300	1,100
1000						
	10,900	5,500	3,600	2,700	2,200	1,800
1250						
	16,200	8,100	5,400	4,100	3,200	2,700
1500						
	22,700	11,400	7,600	5,700	4,500	3,800
1750						
	30,200	15,100	10,100	7,600	6,000	5,000
2000						
	38,700	19,400	12,900	9,700	7,700	6,500
2250						
	48,300	24,200	16,100	12,100	9,700	8,100
2500						
	59,000	29,500	19,700	14,800	11,800	9,800
2750						
	70,900	35,500	23,600	17,700	14,200	11,800
3000						
	83,700	41,900	27,900	20,900	16,800	13,900
3250						
	97,700	48,900	32,600	24,400	19,500	16,300
3500						
	112,700	56,400	37,600	28,200	22,500	18,800
3750						
	128,700	64,400	42,900	32,200	25,700	21,500
4000						
	145,900	73,000	48,600	36,500	29,200	24,300
4250						
	164,200	82,100	54,700	41,100	32,800	27,400
4500						
	183,400	91,700	61,100	45,900	36,700	30,600
4750						
	203,700	101,900	67,900	50,900	40,700	34,000
5000						
	225,200	112,600	75,100	56,300	45,000	37,600
5250						
	247,700	123,900	82,600	61,900	49,500	41,300
5500						
	271,200	135,600	90,400	67,800	54,200	45,200
5750						
	295,900	148,000	98,600	74,000	59,200	49,300
6000						

DETERMINING FIRE FLOW FOR TYPES OF CONSTRUCTION

TYPES OF CONSTRUCTION

For the specific purpose of using the Guide, the following definitions may be used:

Fire-Resistive Construction – Any structure that is considered fire-resistive by any of the four model building codes.

Noncombustible Construction – Any structure having all structural members, including walls, columns, piers, beams, girders, trusses, floors, and roofs of noncombustible material and not qualifying as fire-resistive construction.

Ordinary Construction – Any structure having exterior walls of masonry or other noncombustible material in which the other structural members, including but not limited to columns, floors, roofs, beams, girders, and joists, are wholly or partly of wood or other combustible material.

Heavy timber type buildings are required to satisfy a number of specific provisions (see any of the four model building codes).

Wood Frame Construction – Any structure in which the structural members are wholly or partly of wood or other combustible material and the construction does not qualify as ordinary construction.

OCCUPANCY

Low Hazard Occupancies:

Apartments
Asylums
Churches
Clubs
Colleges and Universities
Dormitories
Dwellings
Hospitals
Hotels
Institutions
Libraries, except Large Stack Room Areas
Museums
Nursing, Convalescent and Care Homes
Office Buildings
Prisons
Public Buildings
Rooming Houses
Schools
Tenements

High Hazard Occupancies:

Aircraft Hangers
Cereal, Feed, Flour and Grist Mills
Chemical Works – High Hazard
Cotton Picker and Opening Operations
Explosives and Pyrotechnics Manufacturing
High-Piled Combustible Storage in excess of 21 feet high
Linoleum and Oilcloth Manufacturing
Linseed Oil Mills
Match Manufacturing
Oil Refineries
Paint Shops
Pyroxylin Plastic Manufacturing and Processing
Shade Cloth Manufacturing
Solvent Extracting
Varnish and Paint Works
Wood Working with Flammable Finishing
Other occupancies involving processing, mixing, storage and dispensing flammable and/or combustible liquids.

Experience has shown that the following credits should normally be applied for the occupancies listed:

Dwellings*, apartments and dormitories	-25%
Hospitals	-20%
Elementary schools	-20%
Junior and Senior high schools	-15%
Open parking garages	-25%

*When applying the standard method.

For other occupancies, good judgment should be used, and the percentage increase or decrease will not necessarily be the same for all buildings that are in the same general category – for example "Colleges and Universities": this could range from a 25% decrease for buildings used only as dormitories to an increase for a chemical laboratory. Even when considering high schools, the decrease should be less if they have extensive shops.

It is expected that in commercial buildings no percentage increase or decrease for occupancy will be applied in most of the fire flow determinations. In general, percentage increase or decrease will not be at the limits of ± 25%.

EXPOSURES

When determining exposures it is necessary to understand that the exposure percentage increase for a fire in a building (x) exposing another building (y) does not necessarily equal the percentage increase when the fire is in building (y) exposing building (x). The Guide gives the maximum possible percentage for exposure at specified distances. However, these maximum percentages should not be used for all exposures at those distances. In each case the percentage applied should reflect the actual conditions but should not exceed the percentage listed.

The maximum percentage for the separations listed generally should be used if the exposed building meets all of the following conditions:

a. Same type or a poorer type of construction than the fire building.

b. Same or greater height than the fire building.

c. Contains unprotected exposed openings.

d. Unsprinklered.

Copy the sample Fire Flow Estimate form on the facing page to record and use the actual figures and conditions found locally. Following the blank sample are several filled-in samples for comparison.

APPENDIX D – SAMPLE FIRE FLOW ESTIMATE CHARTS

FIRE FLOW ESTIMATE

Date ______________

City ______________ State ______________

Eng. ______________

Bound Block or Complex. by streets, etc:

Previous Fire Flow No. ______________

Fire Flow No. ______________

Phantom No. ______________

Route No. ______________

Address (name of occupant if prominent)

Sanborn Vol. ________ Page ________

Type Dist. ______________

Fire Area Considered

Types of Construction: ______________

Ground Floor Area ______________ No. of Stories ______________

Total Floor Area (if needed) ______________

Fire Flow From Table: ______________ gpm(a)

Occupancy: ______________ Add or Subtract ________ % ________

Sub Total ________ gpm(b)

Automatic Sprinklers: ______________ Subtract ________ % x b = - ________

Sub Total ________

Exposures: Distance Exposure

1. Front ______________ Add ________ %
2. Left ______________ ________ %
3. Rear ______________ ________ %
4. Right ______________ ________ %

Total ________ %

Notes and/or Calculations:

Use ________ % x b = + ________

Total ________ gpm

Fire Flow Required ________ gpm

Draw Sketch on other side if needed.

INSURANCE SERVICES OFFICE
TYPICAL BLOCK A

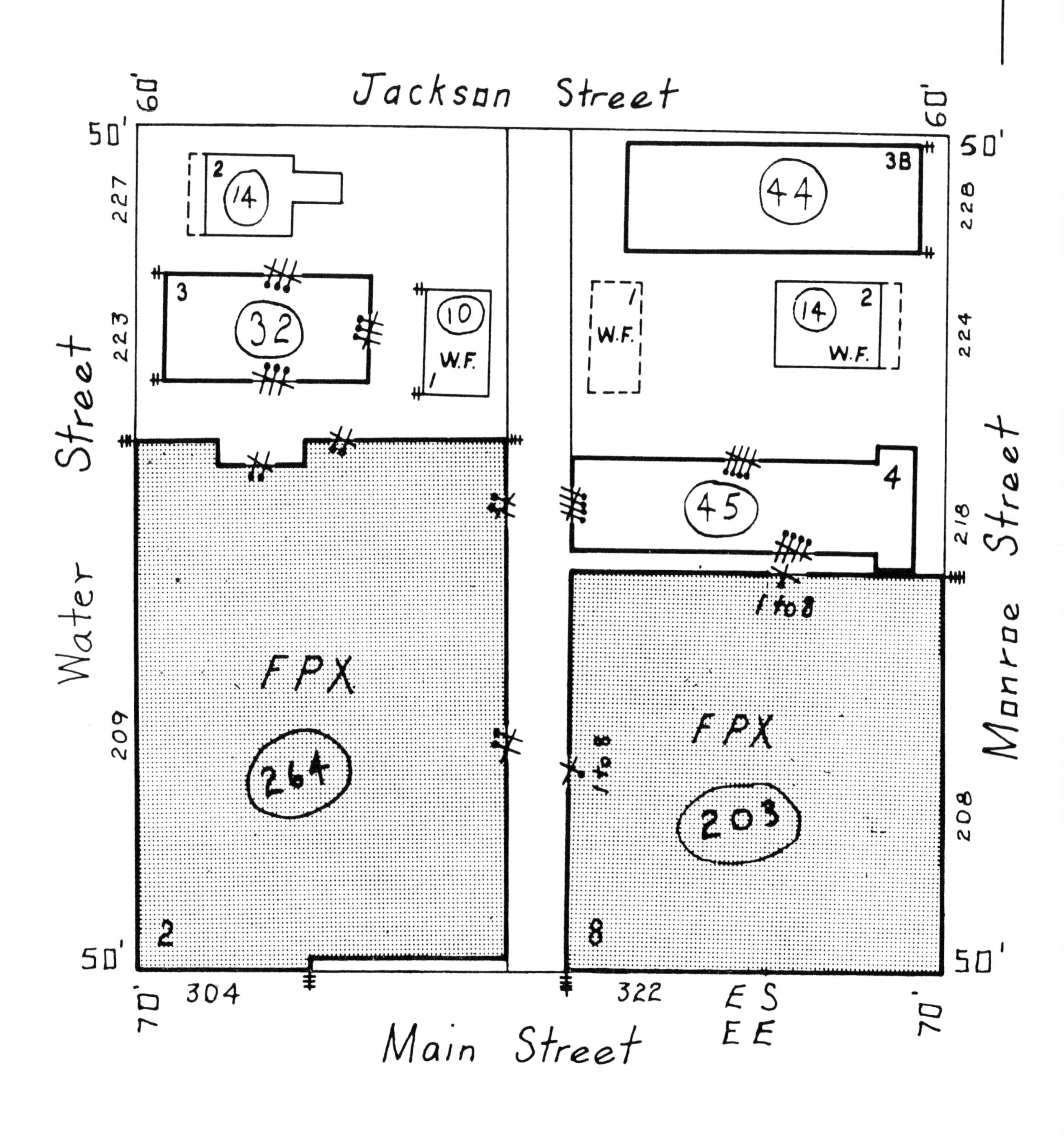

INSURANCE SERVICES OFFICE
MUNICIPAL SURVEY SERVICE
FIRE FLOW ESTIMATE

Date 11/87

City EXAMPLE A State

Eng.

Bound Block or Complex. by streets, etc:

Previous Fire Flow No.

JACKSON, MONROE, MAIN, WATER

Fire Flow No.

Phantom No.

Route No.

Address (name of occupant if prominent)

Sanborn Vol. Page

Type Dist. Business

322 MAIN STREET

Fire Area Considered

Types of Construction: FIRE - RESISTIVE (ES-EE)

Ground Floor Area 20,300 No. of Stories 8 (use 3)

Total Floor Area (if needed)

Fire Flow From Table: 2750 gpm(a)

Occupancy: RETAIL, OFFICES, etc. Add or Subtract 0 % 0

Sub Total 2750 gpm(b)

Automatic Sprinklers: NO Subtract 0 % x b = 0

Sub Total 2750

Exposures:	Distance	Exposure		
1. Front	70'	13-st., F.R.	Add	10 %
2. Left	20'	2-st., F.R.		15 %
3. Rear	5'	4-st., ord.		25 %
4. Right	50'	2-st., ord.		10 %
			Total	60 %

Notes and/or Calculations:

Use 60 % x b = + 1650

Total 4400 gpm

Fire Flow Required 4500 gpm

Draw Sketch on other side if needed.

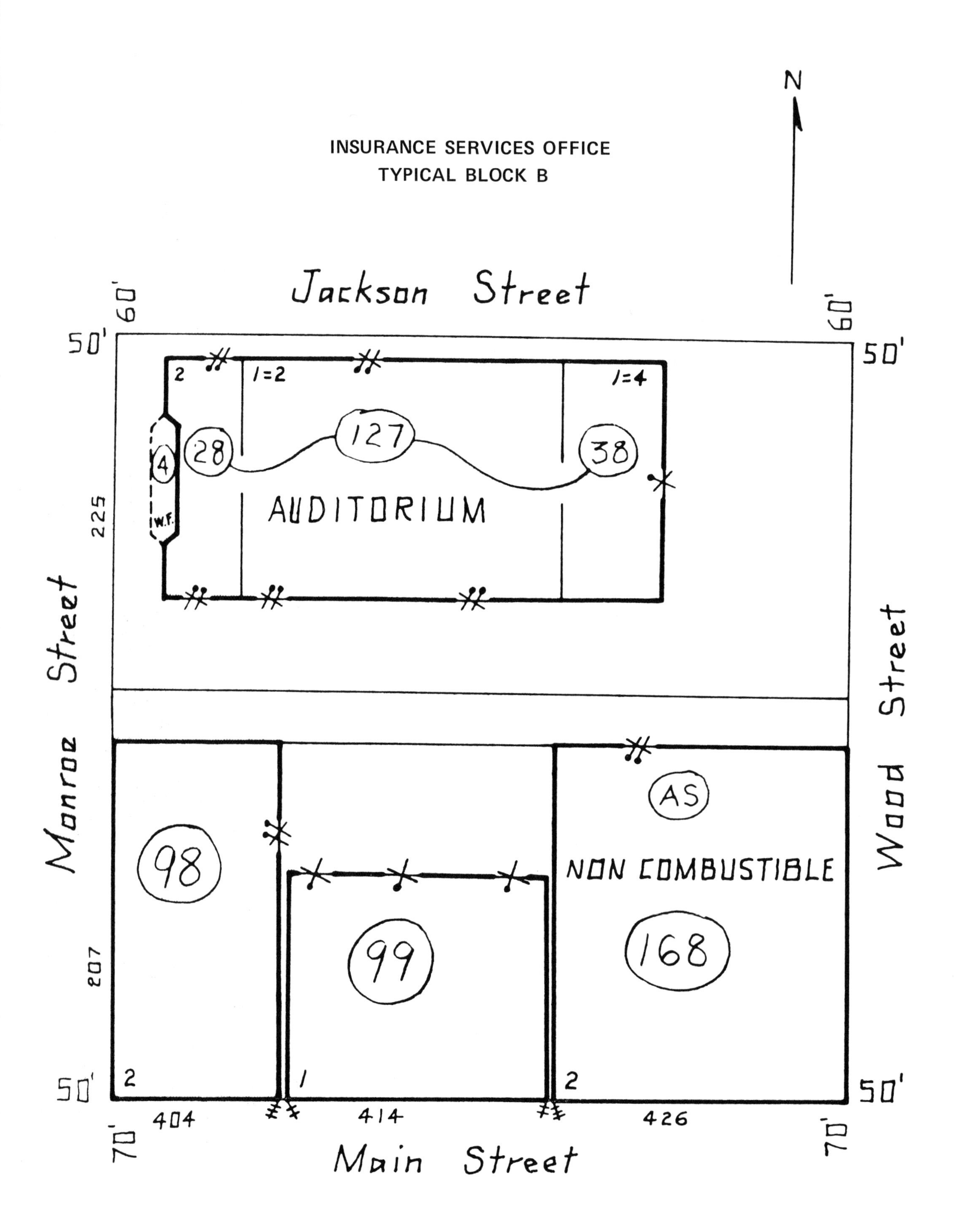
N
INSURANCE SERVICES OFFICE
TYPICAL BLOCK B
Jackson Street
Monroe Street
Wood Street
Main Street
60'
50'
225
207
70'
2
1=2
1=4
4
W.F.
28
127
38
AUDITORIUM
AS
98
99
NON COMBUSTIBLE
168
1
404
414
426

INSURANCE SERVICES OFFICE
MUNICIPAL SURVEY SERVICE
FIRE FLOW ESTIMATE

Date 11/87

City EXAMPLE B State

Eng.

Bound Block or Complex. by streets, etc:

Previous Fire Flow No.

Fire Flow No.

MONROE, JACKSON, WOOD, MAIN

Phantom No.

Route No.

Address (name of occupant if prominent)

Sanborn Vol. Page

Type Dist. Business

225 MONROE STREET, AUDITORIUM

Fire Area Considered

Types of Construction: W.F./ord/ord

Ground Floor Area 400/2,800/16,500 No. of Stories 1/2/1

Total Floor Area (if needed) 22,500

Fire Flow From Table: 2,750* gpm(a)

Occupancy: AUDITORIUM Add or Subtract 0 % 0

Sub Total 2,750 gpm(b)

Automatic Sprinklers: NO Subtract 0 % x b = 0

Sub Total 2,750

Exposures:	Distance	Exposure		
1. Front	75'	3-st. ord, 2-st. W.F.	Add	10 %
2. Left	60'	1-st. ord		10 %
3. Rear	141'	6-st. ord		2 %
4. Right	57'-108'	1+2 st. ord		8 %
			Total	30 %

Use 30 % x b = + 825

Notes and/or Calculations:

$\frac{400}{22{,}500} \times (4000) = 71$

$\frac{22{,}100}{22{,}500} \times (2750) = 2{,}701$

2,772 = 2,750

Total 3,575 gpm

Fire Flow Required 3,500 gpm

Draw Sketch on other side if needed.

INSURANCE SERVICES OFFICE
TYPICAL BLOCK C

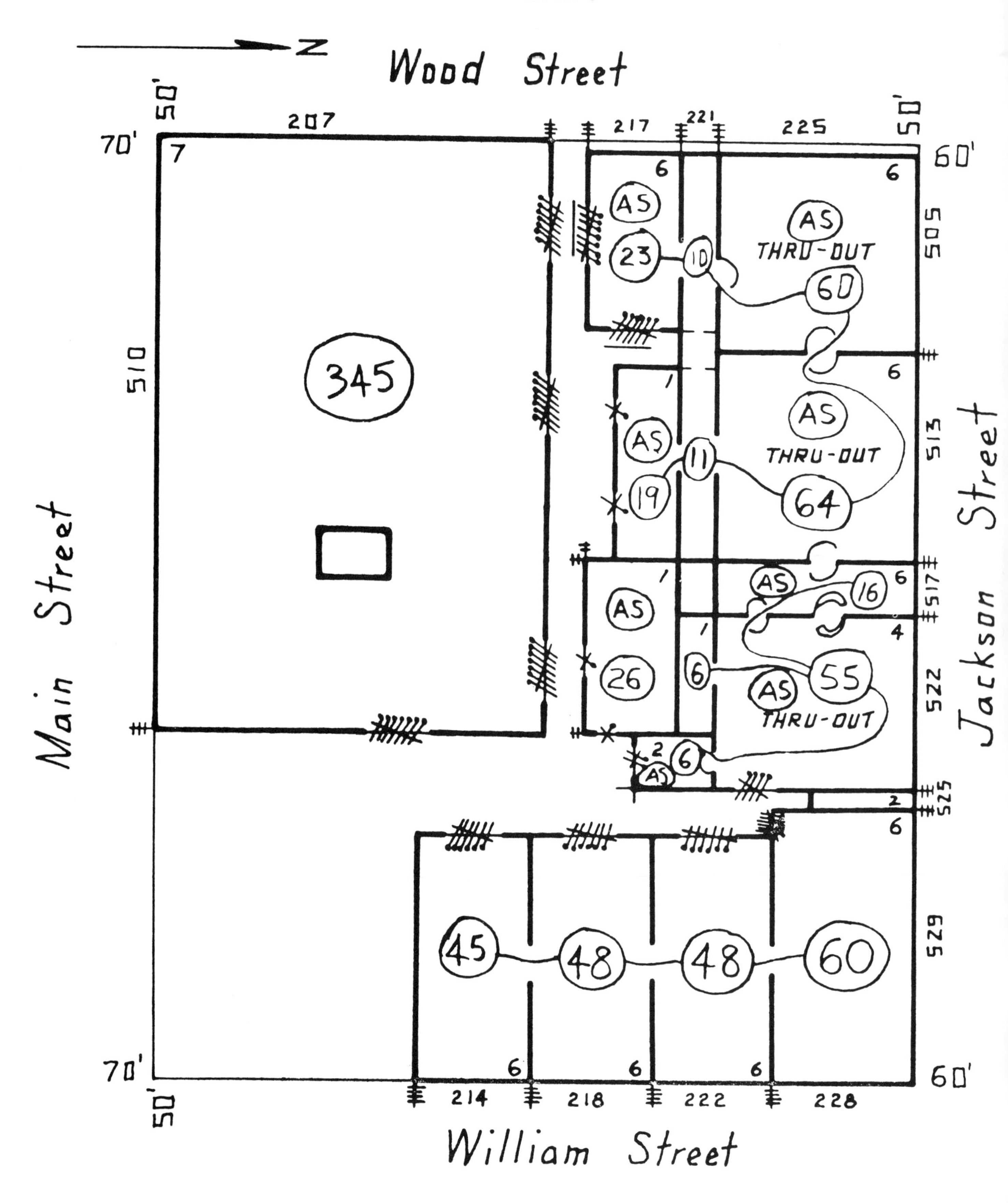

INSURANCE SERVICES OFFICE
MUNICIPAL SURVEY SERVICE
FIRE FLOW ESTIMATE

Date 11/87

City EXAMPLE C State

Eng.

Bound Block or Complex. by streets, etc:

Previous Fire Flow No.

Fire Flow No.

JACKSON, WILLIAM, MAIN, WOOD

Phantom No.

Route No.

Address (name of occupant if prominent)

Sanborn Vol. Page

Type Dist. Business

Fire Area Considered

Types of Construction: ord

Ground Floor Area 34,500 No. of Stories 7

Total Floor Area (if needed) 241,500

Fire Flow From Table: 8,000 gpm(a)

Occupancy: DEPARTMENT STORE Add or Subtract 0 % 0

Sub Total 8,000 gpm(b)

Automatic Sprinklers: NO Subtract 0 % x b = 0

Sub Total 8,000

Exposures:	Distance	Exposure		
1. Front	70'	2-3 st. ord	Add	5 %
2. Left	50'	2 st. ord		7 %
3. Rear	15'-28'	4-6 st. ord		20 %
4. Right	38'	6 st. ord and open		10 %

Total 42 %

Notes and/or Calculations:

Use 42 % x b = + 3,360

Total 11,360 gpm

Fire Flow Required 11,000 gpm

Draw Sketch on other side if needed.

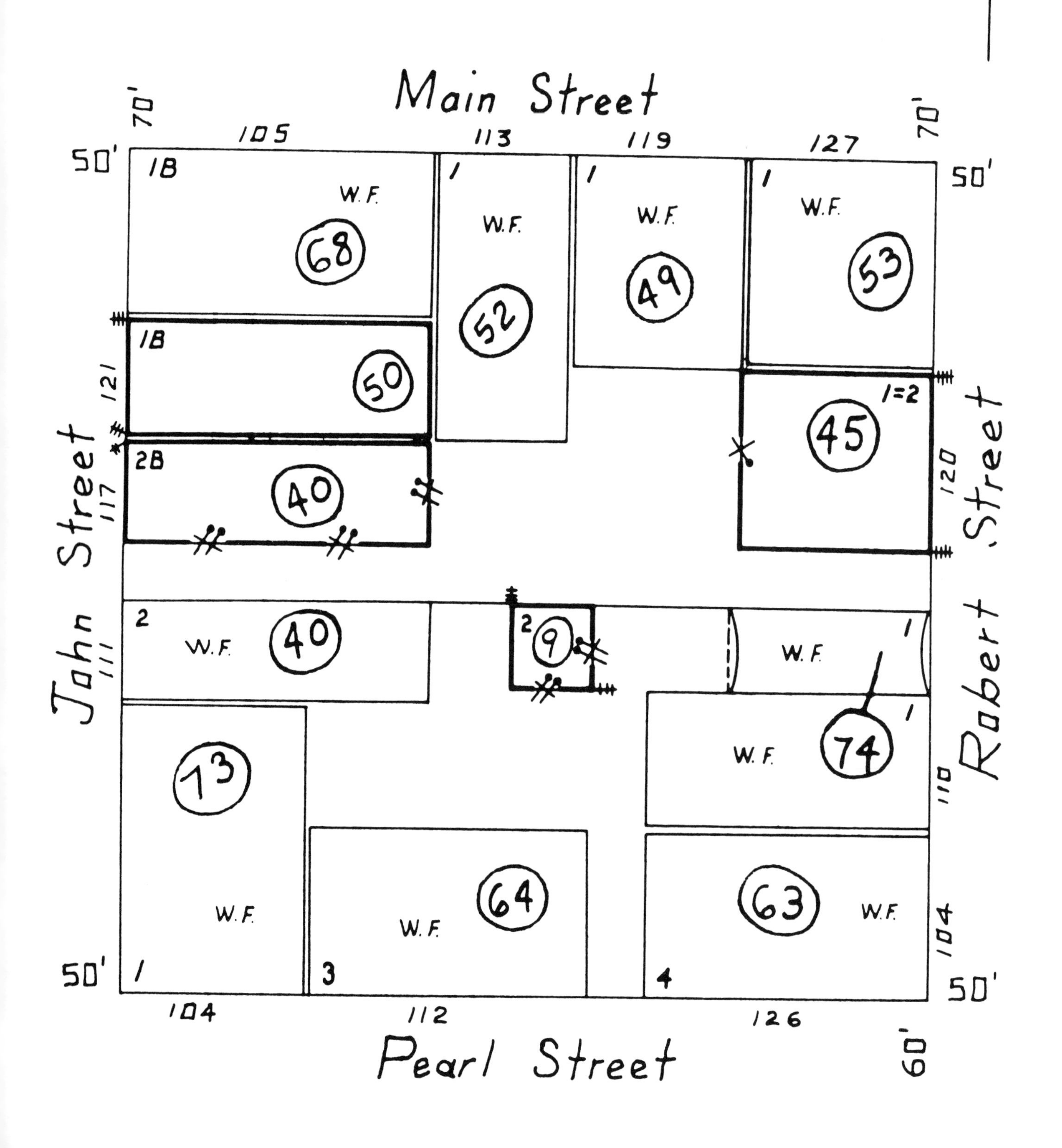
INSURANCE SERVICES OFFICE
TYPICAL BLOCK D
N
Main Street
Pearl Street
John Street
Robert Street
70'
50'
60'
105
113
119
127
104
112
126
121
117
111
120
110
104
1B
2B
W.F.
68
52
49
53
50
45
40
9
74
73
64
63
1=2

INSURANCE SERVICES OFFICE
MUNICIPAL SURVEY SERVICE
FIRE FLOW ESTIMATE

Date 11/87

City EXAMPLE D State

Eng.

Bound Block or Complex. by streets, etc:

Previous Fire Flow No.

Fire Flow No.

MAIN, ROBERT, PEARL, JOHN

Phantom No.

Route No.

Address (name of occupant if prominent)

Sanborn Vol. Page

Type Dist. Business

111 JOHN AND 104 - 112 PEARL

Fire Area Considered

Types of Construction: WOOD FRAME

Ground Floor Area 4,000/7,300/6,400 No. of Stories 2/1/3

Total Floor Area (if needed) 34,500

Fire Flow From Table: 5,000 gpm(a)

Occupancy: RETAIL Add or Subtract 0 % 0

Sub Total 5,000 gpm(b)

Automatic Sprinklers: NO Subtract 0 % x b = 0

Sub Total

Exposures:	Distance	Exposure		
1. Front	60'	2-st. ord	Add	12 %
2. Left	50'	1-st. W.F.		8 %
3. Rear	20'	2-st. ord		18 %
4. Right	20'	4-st. W.F.		20 %
			Total	58 %

Notes and/or Calculations:

Use 58 % x b = + 2,900

Total 7,900 gpm

Fire Flow Required 8,000 gpm

Draw Sketch on other side if needed.

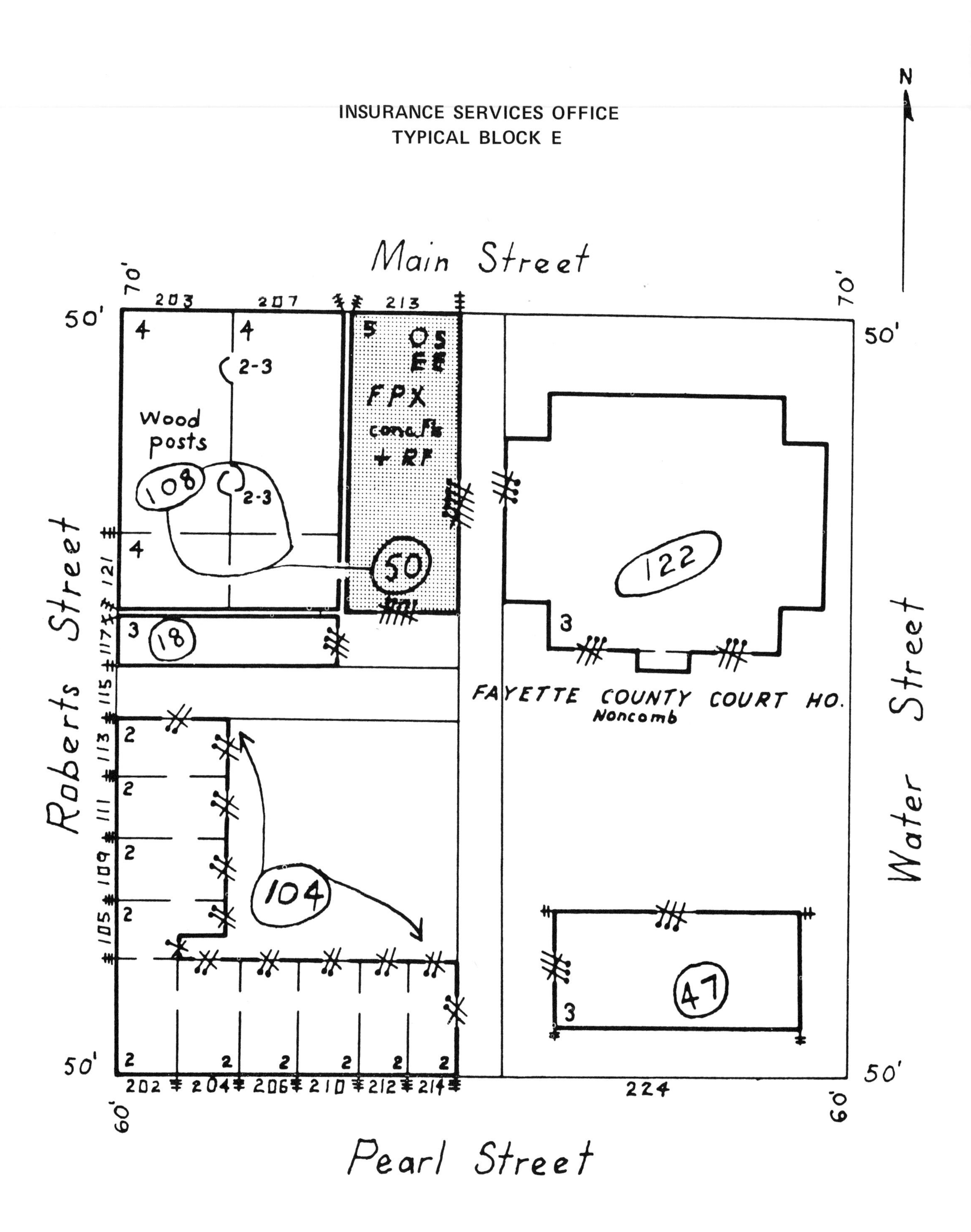

N
INSURANCE SERVICES OFFICE
TYPICAL BLOCK E
Main Street
Roberts Street
Water Street
Pearl Street
70'
70'
50'
50'
50'
50'
60'
60'
203
207
213
4
4
2-3
Wood posts
108
2-3
4
121
5
FPX
50
3
18
117
115
113
111
109
105
2
2
2
2
104
122
3
FAYETTE COUNTY COURT HO.
Noncomb
47
3
202
204
206
210
212
214
224

INSURANCE SERVICES OFFICE
MUNICIPAL SURVEY SERVICE
FIRE FLOW ESTIMATE

Date 11/87

City EXAMPLE E State

Eng.

Bound Block or Complex. by streets, etc:

Previous Fire Flow No.

Fire Flow No.

MAIN, WATER, PEARL, ROBERTS

Phantom No.

Route No.

Address (name of occupant if prominent)

Sanborn Vol. Page

Type Dist. Business

203 - 213 MAIN STREET

Fire Area Considered

Types of Construction: ord/ FIRE-RESISTIVE (os)

Ground Floor Area 10,800/5,000 No. of Stories 4/5

Total Floor Area (if needed) 43,200/25,000 = 68,200

Fire Flow From Table: 4,000* gpm(a)

Occupancy: RETAIL, BANK, OFFICES, etc. Add or Subtract 0 % 0

Sub Total 4,000 gpm(b)

Automatic Sprinklers: NO Subtract – % x b = 0

Sub Total 4,000

Exposures:	Distance	Exposure		
1. Front	70'	2-4 st. ord	Add	8 %
2. Left	20'	3-st. N.C.		18 %
3. Rear	0'	FW - (dia.)		15 %
4. Right	50'	1-st. W.F., 2-st. ord		7 %
			Total	48 %

Use 48 % x b = + 1,920

Notes and/or Calculations:

$\frac{43,200}{68,200} \times 4,750 = 3,010$

$\frac{25,000}{68,200} \times 2,750 = 1,010$

4,020 = 4000*

Total 5,920 gpm

Fire Flow Required 6,000 gpm

Draw Sketch on other side if needed.

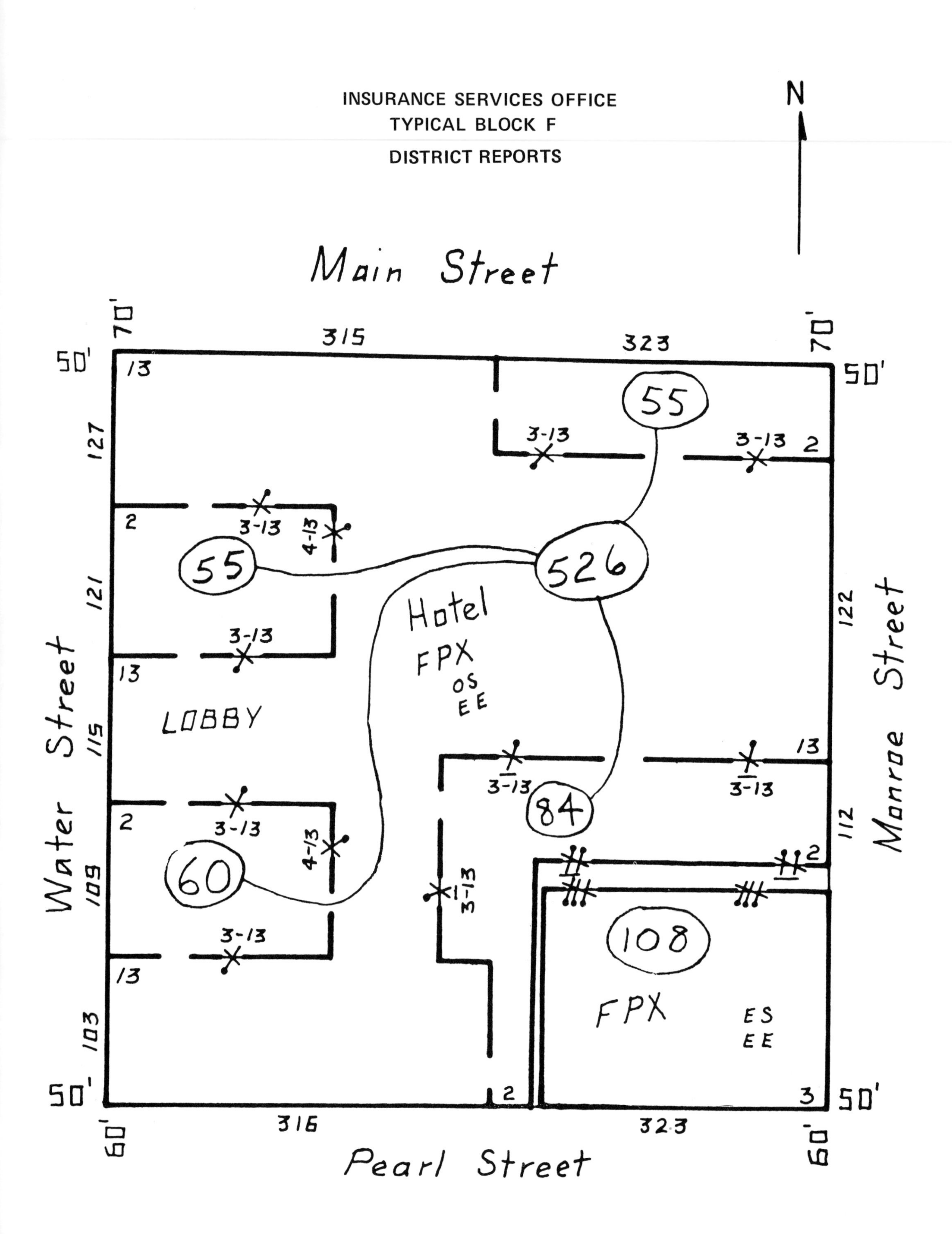
INSURANCE SERVICES OFFICE
TYPICAL BLOCK F
DISTRICT REPORTS
N
Main Street
Pearl Street
Water Street
Monroe Street
315
323
316
323
Hotel
FPX
OS
EE
LOBBY
FPX
ES
EE
55
55
526
60
84
108
3-13
4-13
1
3-13
70'
50'
60'
127
121
115
109
103
122
112

INSURANCE SERVICES OFFICE
MUNICIPAL SURVEY SERVICE
FIRE FLOW ESTIMATE

Date 11/87

City EXAMPLE F State

Eng.

Bound Block or Complex. by streets, etc:

Previous Fire Flow No.

Fire Flow No.

MAIN, MONROE, PEARL, WATER

Phantom No.

Route No.

Address (name of occupant if prominent)

Sanborn Vol. Page

Type Dist. Business

315 MAIN STREET

Fire Area Considered

Types of Construction: FIRE-RESISTIVE

Ground Floor Area 25,400/52,600 No. of Stories 2/13 O.S. (use 6)

Total Floor Area (if needed) 366,400

Fire Flow From Table: 6,000 gpm(a)

Occupancy: HOTEL Add or (Subtract) 10 % 600

Sub Total 5,400 gpm(b)

Automatic Sprinklers: NO Subtract 0 % x b = 0

Sub Total 5,400

Exposures: Distance Exposure

1. Front 70' 2-8-st. F.R. Add 5 %
2. Left 50' 1-st. ord 6 %
3. Rear 10'-60' 6-st. F.R., 2-st. ord 15 %
4. Right 60' 3-st. N.C., 3-st. ord 8 %

Total 34 %

Notes and/or Calculations: Use 34 % x b = + 1,836

Total 7,236 gpm

Fire Flow Required 7,000 gpm

Draw Sketch on other side if needed.

INSURANCE SERVICES OFFICE
TYPICAL BLOCK G

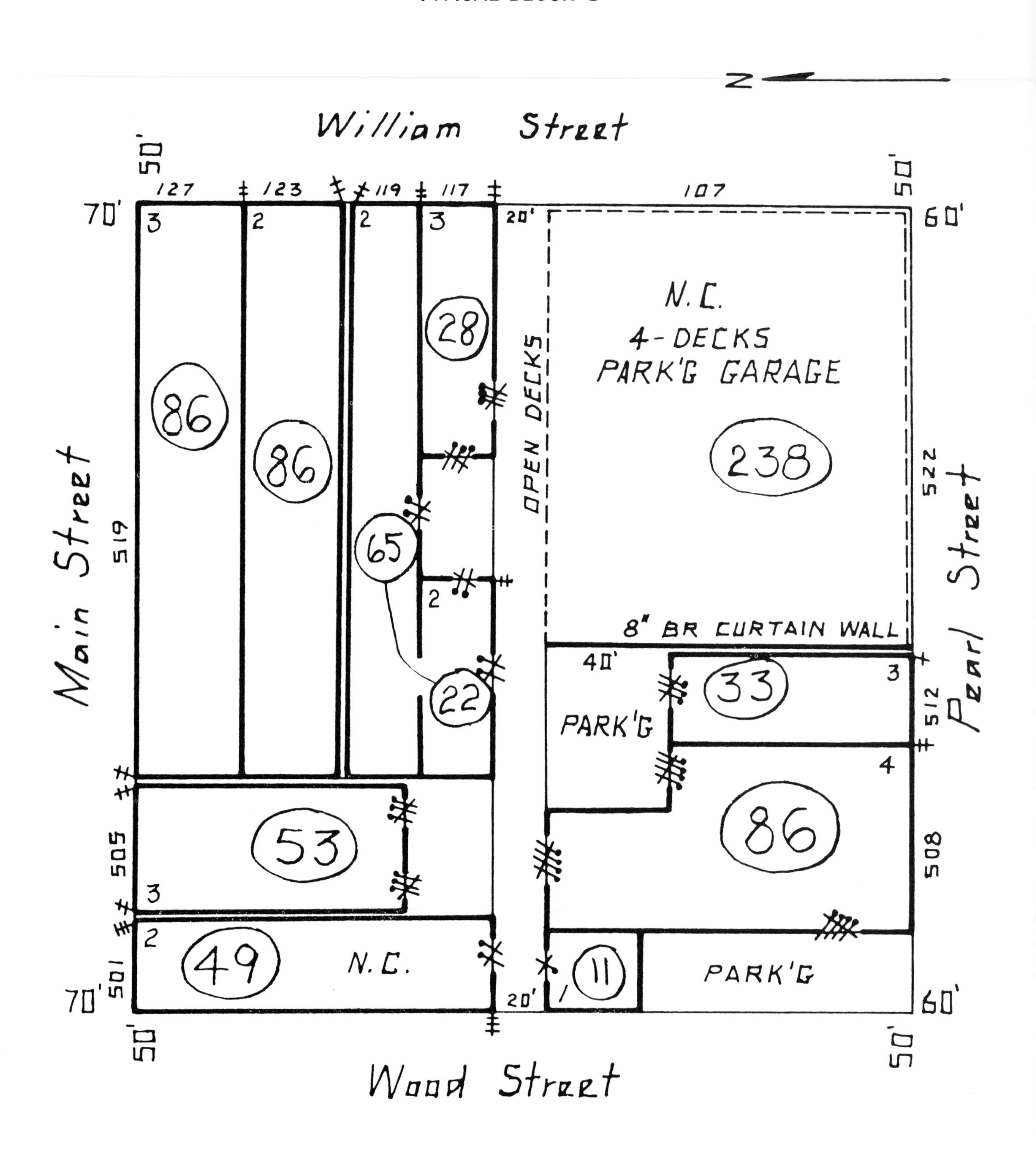

INSURANCE SERVICES OFFICE
MUNICIPAL SURVEY SERVICE
FIRE FLOW ESTIMATE

Date 11/87

City EXAMPLE G State

Eng.

Bound Block or Complex. by streets, etc:

Previous Fire Flow No.

Fire Flow No.

MAIN, WILLIAM, PEARL, WOOD

Phantom No.

Route No.

Address (name of occupant if prominent)

Sanborn Vol. Page

Type Dist. Business

522 PEARL STREET

Fire Area Considered

Types of Construction: N.C.

Ground Floor Area 23,800 No. of Stories 4

Total Floor Area (if needed)

Fire Flow From Table: 4,500 gpm(a)

Occupancy: PARKING GARAGE Add or (Subtract) 25 % 1,125

Sub Total 3,375 gpm(b)

Automatic Sprinklers: NO Subtract 0 % x b = 0

Sub Total 3,375

Exposures:	Distance	Exposure		
1. Front	60'	1-st ord	Add	5 %
2. Left	0'	F.W.		0 %
3. Rear	20'	open to 2-3st. ord		20 %
4. Right	50'	1 & 2-st. ord		10 %
			Total	35 %

Notes and/or Calculations:

Use 35 % x b = + 1,181

Total 4,556 gpm

Fire Flow Required 4,500 gpm

Draw Sketch on other side if needed.

INSURANCE SERVICES OFFICE
TYPICAL BLOCK H

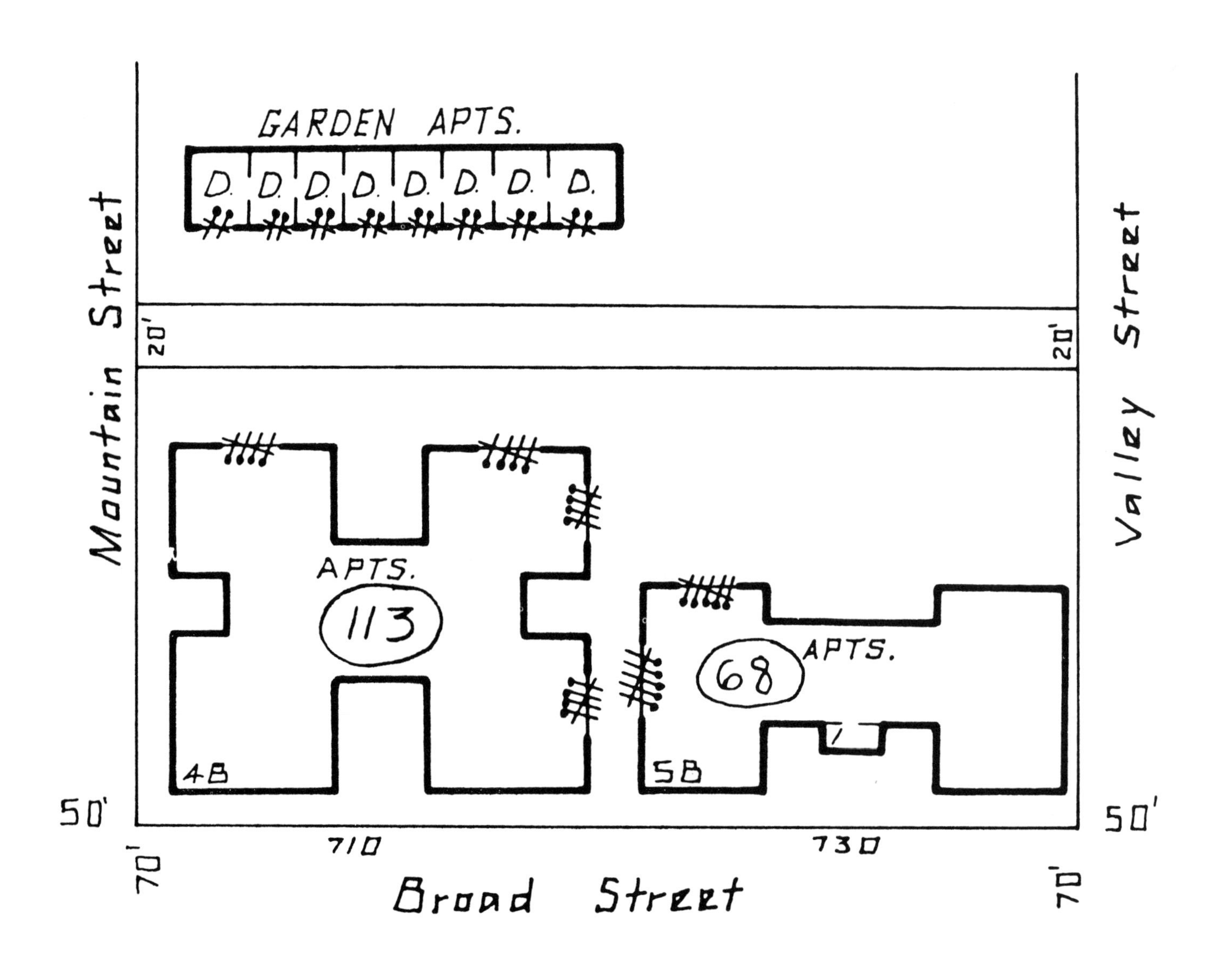

INSURANCE SERVICES OFFICE
MUNICIPAL SURVEY SERVICE
FIRE FLOW ESTIMATE

Date 11/87

City EXAMPLE H State ______

Eng. ______

Bound Block or Complex. by streets, etc:

Previous Fire Flow No. ______

Fire Flow No. ______

BROAD ST. bet. MOUNTAIN AND VALLEY

Phantom No. ______

Route No. ______

Address (name of occupant if prominent)

Sanborn Vol. ______ Page ______

Type Dist. Apartment

710 BROAD STREET

Fire Area Considered

Types of Construction: ord

Ground Floor Area 11,300 No. of Stories 4

Total Floor Area (if needed) ______

Fire Flow From Table: ______ 3,750 gpm(a)

Occupancy: APARTMENTS Add or Subtract 25 % 938

Sub Total 2,812 gpm(b)

Automatic Sprinklers: NO Subtract 0 % x b = 0

Sub Total ______

Exposures:	Distance	Exposure		
1. Front	90'	4-st. ord.	Add	10 %
2. Left	70'	3-st. ord.		8 %
3. Rear	65'	2-st. ord.		5 %
4. Right	15'	5-st. ord.		20 %
			Total	43 %

Notes and/or Calculations:

Use 43 % x b = + 1,209

Total 4,021 gpm

Fire Flow Required 4,000 gpm

Draw Sketch on other side if needed.

INSURANCE SERVICES OFFICE
TYPICAL BLOCK I

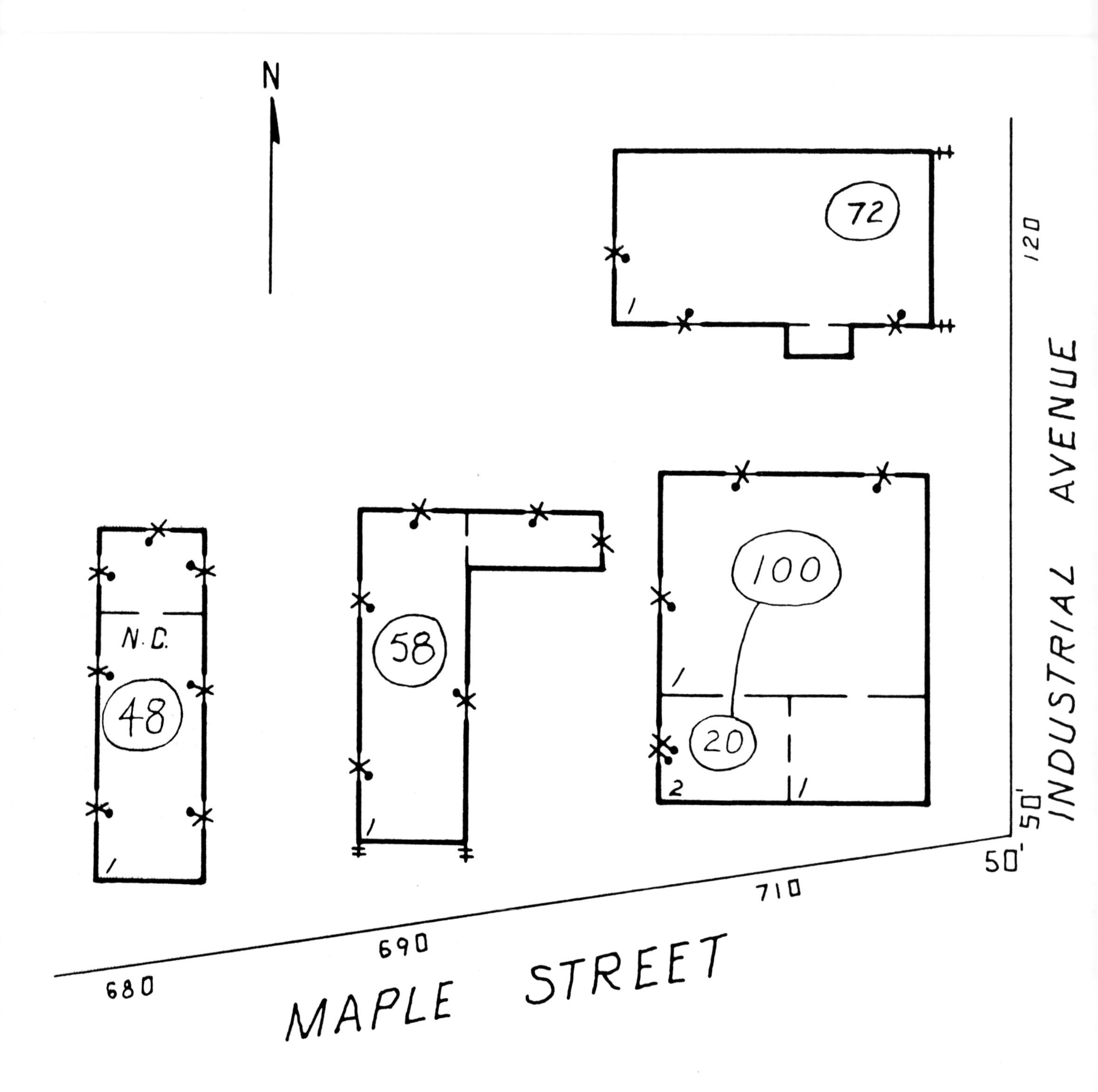

INSURANCE SERVICES OFFICE
MUNICIPAL SURVEY SERVICE
FIRE FLOW ESTIMATE

Date 11/87

City EXAMPLE I State ____

Eng. ____

Bound Block or Complex. by streets, etc:

Previous Fire Flow No. ____

N.W. cor. MAPLE ST. and INDUSTRIAL AVE.

Fire Flow No. ____

Phantom No. ____

Route No. ____

Address (name of occupant if prominent)

Sanborn Vol. ____ Page ____

Type Dist. Industrial

710 MAPLE ST.

Fire Area Considered

Types of Construction: ord/ord

Ground Floor Area 2,000/10,000 No. of Stories 2/1

Total Floor Area (if needed) 14,000

Fire Flow From Table: 2,250 gpm(a)

Occupancy: TIRE WAREHOUSE Add or Subtract 10 % 225

Sub Total 2,475 gpm(b)

Automatic Sprinklers: NO Subtract 0 % x b = 0

Sub Total 2,475

Exposures:	Distance	Exposure		
1. Front	80'	1-st. ord. ind	Add	10 %
2. Left	20'-70'	1-st. ord.		15 %
3. Rear	50'	1-st. ord. ind		15 %
4. Right	110'	1-st. ord. ind		5 %
			Total	45 %

Notes and/or Calculations:

Use 45 % x b = + 1,114

Total 3,589 gpm

Fire Flow Required 3,500 gpm

Draw Sketch on other side if needed.

INSURANCE SERVICES OFFICE
TYPICAL BLOCK J

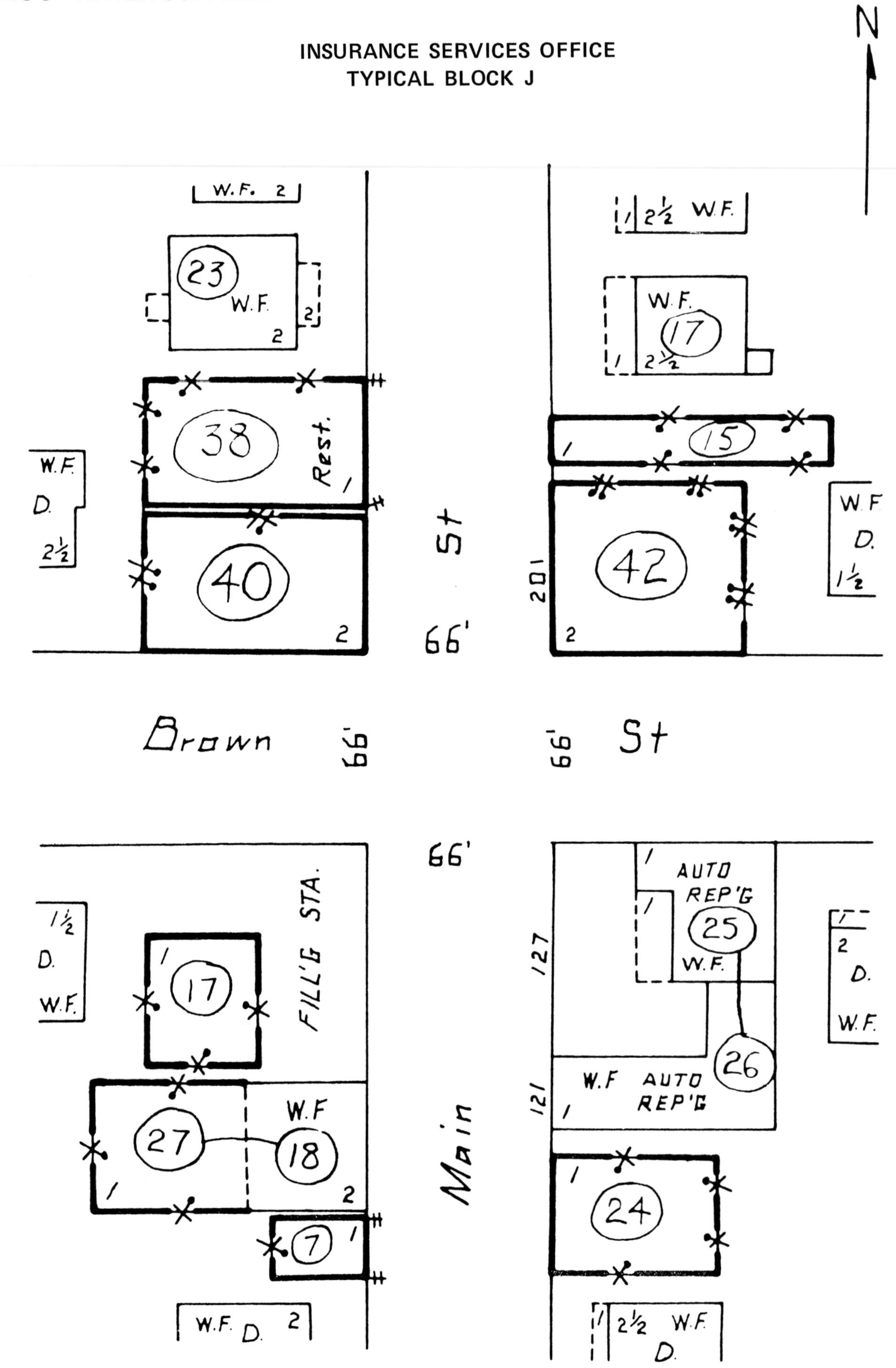

INSURANCE SERVICES OFFICE
MUNICIPAL SURVEY SERVICE
FIRE FLOW ESTIMATE

Date 11/87

City EXAMPLE J State ______

Eng. ______

Bound Block or Complex. by streets, etc:

S.E. corner MAIN and BROWN

Previous Fire Flow No. ______

Fire Flow No. ______

Phantom No. ______

Route No. ______

Address (name of occupant if prominent)

121-127 MAIN ST.

Sanborn Vol. ______ Page ______

Type Dist. Business

Fire Area Considered

Types of Construction: W.F./W.F.

Ground Floor Area 2,500/2,600 No. of Stories 1/1

Total Floor Area (if needed) 5,100

Fire Flow From Table: 2,000 gpm(a)

Occupancy: AUTO REPAIR Add or Subtract 10 % +200

Sub Total 2200 gpm(b)

Automatic Sprinklers: No Subtract 0 % x b = 0

Sub Total 2200

Exposures:	Distance	Exposure		
1. Front	66' - 140'	2-st. W. F. & 1st Ord.	Add	8 %
2. Left	66'	2-st. Ord.		10 %
3. Rear	20'	2-st. W. F.		20 %
4. Right	10'	1-st. Ord.		15 %
			Total	53 %

Notes and/or Calculations:

Use 53 % x b = + 1166

Total 3366 gpm

Fire Flow Required 3500 gpm

Draw Sketch on other side if needed.

Appendix C
ISO *Fire Suppression Rating Schedule:* Needed Fire Flow

NEEDED FIRE FLOW

300. GENERAL:

This item develops Needed Fire Flows for selected locations throughout the city which are used in the review of subsequent items of this Schedule. The calculation of a Needed Fire Flow (NFF_i) for a subject building in gallons per minute (gpm) considers the Construction (C_i), Occupancy (O_i), Exposure (X_i) and Communication (P_i) of each selected building, or fire division, as outlined below.

310. CONSTRUCTION FACTOR (C_i):

That portion of the Needed Fire Flow attributed to the construction and area of the selected building is determined by the following formula:

$C_i = 18F(A_i)^{0.5}$

F = Coefficient related to the class of construction:

F = 1.5 for Construction Class 1* (Frame)
= 1.0 for Construction Class 2* (Joisted Masonry)
= 0.8 for Construction Class 3* (Non-Combustible) and Construction Class 4* (Masonry Non-Combustible)
= 0.6 for Construction Class 5* (Modified Fire Resistive) and Construction Class 6* (Fire Resistive)

A_i = Effective* area

In buildings with mixed construction a value, C_{im}, shall be calculated for each class of construction using the effective area of the building. These C_{im} values are multiplied by their individual percentage of the total area. The C_i applicable to the entire building is the sum of these values. However, the value of the C_i shall not be less than the value for any part of the building based upon its own construction and area.

The maximum value of C_i is limited by the following:

8,000 gpm for Construction Classes 1 and 2
6,000 gpm for Construction Classes 3, 4, 5 and 6
6,000 gpm for a 1-story building of any class of construction.

The minimum value of C_i is 500 gpm. The calculated value of C_i shall be rounded to the nearest 250 gpm.

320. OCCUPANCY FACTOR (O_i):

The factors below reflect the influence of the occupancy in the selected building on the Needed Fire Flow.

Occupancy Combustibility Class*	Occupancy Factor (O_i)
C-1*(Non-Combustible)	0.75
C-2*(Limited Combustible)	0.85
C-3*(Combustible)	1.00
C-4*(Free Burning)	1.15
C-5*(Rapid Burning)	1.25

330. EXPOSURES (X_i) AND COMMUNICATION (P_i) FACTORS:

The factors developed in this item reflect the influence of exposed and communicating buildings on the Needed Fire Flow. A value for $(X_i + P_i)$ shall be developed for each side of the subject building:

$(X+P)_i = 1.0 + \sum_{i=1}^{n} (X_i + P_i)$, maximum 1.75, where n = number of sides of subject building.

A. **Factor for Exposure (X_i):**

The factor for X_i depends upon the construction and length-height value* (length of wall in feet, times height in stories) of the exposed building and the distance between facing walls of the subject building and the exposed building, and shall be selected from Table 330.A.

*When an asterisk is shown next to a term in this item, the term is defined in greater detail in the Commercial Fire Rating Schedule.

TABLE 330.A
FACTOR FOR EXPOSURE (X_i)

Construction of Facing Wall of Subject Bldg.	Distance Feet to the Exposed Building	Length-Height of Facing Wall of Exposed Building	Construction of Facing Wall of Exposed Building Classes			
			1,3	2, 4, 5, & 6		
				Unprotected Openings	Semi-Protected Openings (wired glass or outside open sprinklers)	Blank Wall
Frame, Metal or Masonry with Openings	0-10	1-100	0.22	0.21	0.16	0
		101-200	0.23	0.22	0.17	0
		201-300	0.24	0.23	0.18	0
		301-400	0.25	0.24	0.19	0
		Over 400	0.25	0.25	0.20	0
	11-30	1-100	0.17	0.15	0.11	0
		101-200	0.18	0.16	0.12	0
		201-300	0.19	0.18	0.14	0
		301-400	0.20	0.19	0.15	0
		Over 400	0.20	0.19	0.15	0
	31-60	1-100	0.12	0.10	0.07	0
		101-200	0.13	0.11	0.08	0
		201-300	0.14	0.13	0.10	0
		301-400	0.15	0.14	0.11	0
		Over 400	0.15	0.15	0.12	0
	61-100	1-100	0.08	0.06	0.04	0
		101-200	0.08	0.07	0.05	0
		201-300	0.09	0.08	0.06	0
		301-400	0.10	0.09	0.07	0
		Over 400	0.10	0.10	0.08	0
Blank Masonry Wall	Facing Wall of the Exposed Building Is Higher Than Subject Building: Use the above table EXCEPT use only the Length-Height of Facing Wall of the Exposed Building ABOVE the height of the Facing Wall of the Subject Building. Buildings five stories or over in height, consider as five stories.					
	When the Height of the Facing Wall of the Exposed Building is the Same or Lower than the Height of the Facing Wall of the Subject Building, $X_i = 0$.					

330. EXPOSURE (X_i) AND COMMUNICATION (P_i) FACTORS: (Continued)

B. **Factor for Communications (P_i):**

The factor for P_i depends upon the protection for communicating party wall* openings and the length and construction of communications between fire divisions* and shall be selected from Table 330.B. When more than one communication type exists in any one side wall, apply only the largest factor P_i for that side. When there is no communication on a side, $P_i = 0$.

*When an asterisk is shown next to a term in this item, the term is defined in greater detail in the Commercial Fire Rating Schedule.

TABLE 330.B
FACTOR FOR COMMUNICATIONS (P_i)

Description of Protection of Passageway Openings	Fire Resistive, Non-Combustible or Slow-Burning Communications				Communications With Combustible Construction					
	Open	Enclosed			Open			Enclosed		
	Any Length	10 Ft. or Less	11 Ft. to 20 Ft.	21 Ft. to 50 Ft. +	10 Ft. or Less	11 Ft. to 20 Ft.	21 Ft. to 50 Ft. +	10 Ft. or Less	11 Ft. to 20 Ft.	21 Ft. to 50 Ft. +
Unprotected	0	+ +	0.30	0.20	0.30	0.20	0.10	+ +	+ +	0.30
Single Class A Fire Door at One End of Passageway	0	0.20	0.10	0	0.20	0.15	0	0.30	0.20	0.10
Single Class B Fire Door at One End of Passageway	0	0.30	0.20	0.10	0.25	0.20	0.10	0.35	0.25	0.15
Single Class A Fire Door at Each End or Double Class A Fire Doors at One End of Passageway	0	0	0	0	0	0	0	0	0	0
Single Class B Fire Door at Each End or Double Class B Fire Doors at One End of Passageway	0	0.10	0.05	0	0	0	0	0.15	0.10	0

+ For over 50 feet, $P_i = 0$.

+ + For unprotected passageways of this length, consider the 2 buildings as a single Fire Division.

Note: When a party wall has communicating openings protected by a single automatic or self-closing Class B fire door, it qualifies as a division wall* for reduction of area.

Note: Where communications are protected by a recognized water curtain, the value of P_i is O.

*When an asterisk is shown next to a term in this item, the term is defined in greater detail in the Commercial Fire Rating Schedule.

340. CALCULATION OF NEEDED FIRE FLOW (NFF_i):

$$NFF_i = (C_i)(O_i)(X + P)_i$$

When a wood shingle roof covering on the building being considered, or on exposed buildings, can contribute to spreading fires add 500 gpm to the Needed Fire Flow.

The Needed Fire Flow shall not exceed 12,000 gpm nor be less than 500 gpm.

The Needed Fire Flow shall be rounded off to the nearest 250 gpm if less than 2500 gpm and to the nearest 500 gpm if greater than 2500 gpm.

Note 1: For 1- and 2-family dwellings not exceeding 2 stories in height, the following Needed Fire Flows shall be used.

Distance between buildings	Needed Fire Flow
Over 100'	500 gpm
31-100'	750
11-30'	1000
10' or less	1500

Note 2: Other habitational buildings, up to 3500 gpm maximum.

Standard Operating Procedures

The following information may be used as the basis for forming local standard operating procedures on placement of fire apparatus at various types of alarms. The material used in this section was developed by Managing Editor Gene P. Carlson, during his tenure as an instructor for the Maryland Fire and Rescue Institute.

POSITIONING OF ENGINE COMPANIES AND TANKERS FOR RURAL ALARMS

For Short Driveways

1. On locations with short driveways, generally it is best to stay on the main road with the first engine just beyond the fire building. This will give tank vehicles room to work on the hard road and use dump valves to portable tanks. If the engine pulls in the drive, there may be less room to operate, and volume pumps and large hose will be necessary to quickly unload the tankers.
2. The second engine on larger structures takes the proper position at the rear of the fire building for fire attack and exposure protection.

NOTE: Sufficient tankers will be necessary to supply the pumpers on both sides of the fire.

For Medium Length Driveways

1. The first engine
   a. If the first engine will be making a reverse lay to a water source, it should be backed in the drive.
   b. If the first engine is a front mount and will be drafting out of a portable tank, it should be backed in so attack lines are close to the building and the suction is headed out, thus reducing positioning problems for tankers with rear dump valves. This operation is also good for midship pumpers in small farmyards.
   c. If the first engine is laying a line up the drive, it must be driven straight in. It is best to position it upwind, before or past the building, allowing room for proper positioning of the equipment truck.
   d. On rural alarms with large farmyards or working areas, usually two or three tankers can dump at the same time if the tankers are equipped with rear dump valves. Tankers equipped with side dumps can usually only be unloaded one or two at a time. If tankers are pumping off, numerous units can be utilized simultaneously, if sufficient portable tanks and fillers are provided. The first engine in these situations can pull in and position itself as necessary, generally just past the fire building.
2. The second engine functioning as a water supply engine may:
   a. Make a reverse lay from the attack engine to a water source, if available.
   b. If the first engine did not lay a supply line, the second engine may back in and make a reverse lay from the attack pumper to the main road and pump water from the tanker unloading site to the attack engine.
   c. If the first engine laid a supply line, the second engine should connect to it and pump water from the tanker unloading site to the fireground.
   d. Function as a tanker hauling water.

For Long Driveways

1. The first engine

   a. It is best for the first engine to lay a supply line up the drive with a clappered siamese at the main road. This will alleviate crowding of the farmyard and traffic problems on the driveway.

   b. For drives longer than the amount of hose carried by the first engine, the first engine should drive in and take the best possible position for fire attack, taking into consideration the following factors:

      - Wind
      - Exposures
      - Working room for hoselines and portable tank
      - Possible unloading of tankers
      - Maintaining access to building for the equipment vehicle

   c. Another option is to have the first engine lay out its hose bed while coming down the driveway so that the hose can be used in a relay to be finished by later-arriving units.

2. The second engine, functioning as a water supply engine may

   a. Connect to the supply line of the first engine and pump water up the drive to the attack pumper. A good position is to back into the drive from the main road and run tanker discharge lines to each side of the second engine, or to draft from the portable tank off the front corner of the engine.

   b. Back down the drive and initiate a relay operation with other responding engines from the attack pumper to a water source or a tanker unloading source at the main road.

   c. Pick up the supply line laid by the first engine for a relay and assist in completing the operation.

   d. Function as a tanker hauling water.

For Tankers

1. Pumping water from tank

   a. Up a lane

      - To avoid congestion on the main road, pull past the lane and then back into the lane aside the supply hose to the fireground, connect and unload. This eliminates backing onto a busy road. If the lane is wide enough, two tankers may be placed side by side.
      - For multiple unloading where the lane is narrow, tankers can pull past the lane, back into it, and then pull onto the main road and take a position on either side of the lane. Discharge lines are then placed into the clappered siamese in the lane or a pumper.

   b. In the yard
   Tankers in the yard must take a position away from fire fighting activities. Consideration must be given to proximity to the fire building, wind direction, terrain, blocking access of other apparatus, and space for turning around. Proper internal tank-to-pump piping and a volume pump are necessary for fast unloading of tankers in any pumping situation.

   The exact position will depend on the operation. Many times it will be advantageous to back in:

      - When pumping directly to a gated intake, tank fill opening, or filling a portable tank with a filler device, the tanker can be positioned some distance from the destination. Hose should be stretched in anticipation of the tanker's arrival. A preconnected tank discharge line will speed the connection process.

- When pumping directly into a portable tank, tankers must be positioned adjacent to the tank. Using a stream shaper or a hard sleeve on the discharge will increase the efficiency of the process.

2. Dumping water from the tank
   a. Using rear dump valves
   Tanker position will depend upon the location of the portable tank and the access to it. Consideration must be given to tanker unloading operations when placing a portable tank. Ideally, three sides should be available for tankers equipped with rear dump valves. This requires a large area.
   b. Using side dump
   The location of the portable tank will again be important since it is necessary for side dumping tankers to drive or back alongside the portable tank. Poor positioning will limit dumping to one tanker at a time.

The following evolutions give specific situations and recommend procedures for dealing with them. All hose sizes specified in this section should be considered as the minimum for the given situation and may be increased depending on local policy. In order to understand the accompanying diagrams, standard symbols are used to denote the various types of apparatus and other features (Figure D.1).

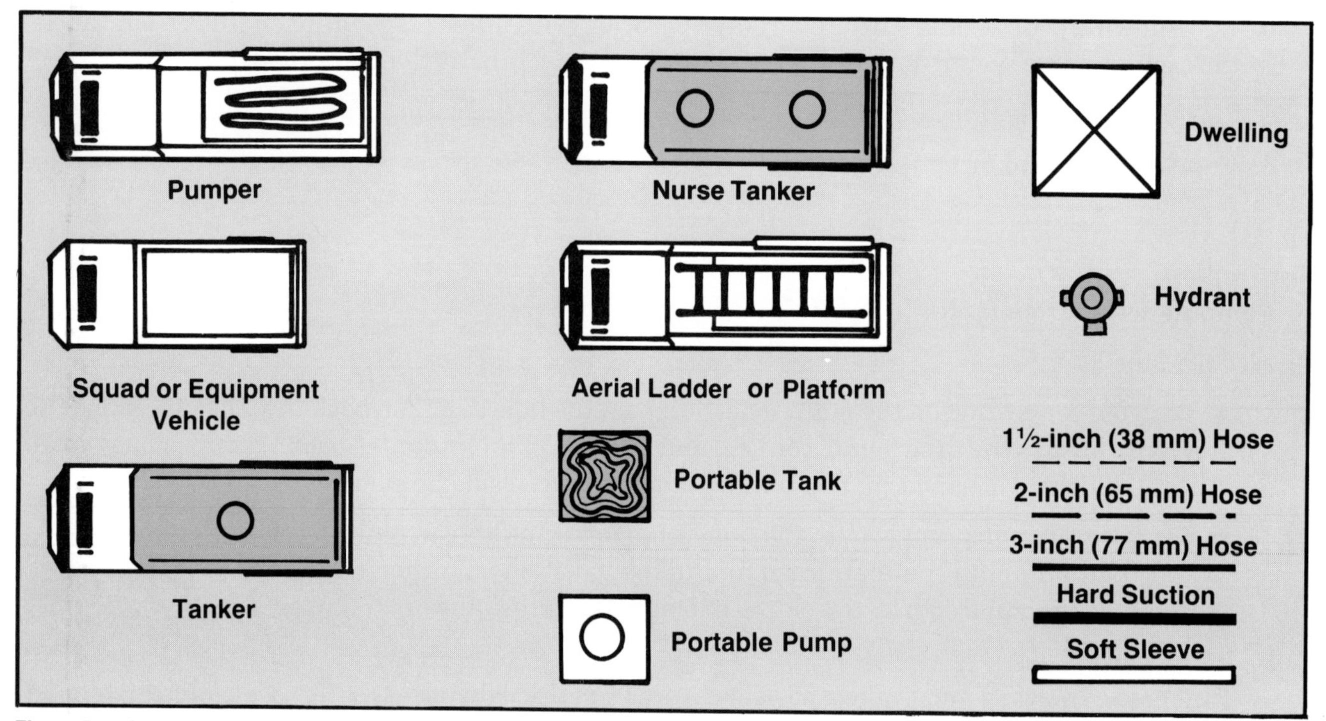

Figure D.1 Standard map symbols used to denote apparatus and equipment used during placement of fire apparatus.

Evolution 1: Small Rural Fire

Response: 1st Alarm — Engine, Engine or Tanker, Squad or Equipment Vehicle
2nd Alarm — Engine or Tanker
2nd Fire — Engine, Tanker, Squad, or Equipment Vehicle

The first engine immediately attacks the fire with 1½-inch (38 mm) or larger preconnected lines from its tank. Additional water will be supplied by the second engine and/or a tanker.

Figure D.2 shows two ways of supplying the first responding engine. One way is to use the 2½-inch (65 mm) gated intake at the pump and the second a 2½-inch (65 mm) tank fill opening. If later responding units are equipped with pumps, the water can be pumped into the first engine; if not equipped with a pump, the engine can draft the water off.

Evolution 2: Large Rural Fire

Response: 1st Alarm — Engine, Engine and Tanker or Two Tankers, and an Equipment Vehicle
2nd Alarm — Engine, Two Tankers
2nd Fire — Tanker, Squad or Equipment Vehicle

This evolution can only be used where there is sufficient room for all apparatus to maneuver.

The first engine attacks the fire using water from its tank with a 2½-inch (65 mm) preconnect or several smaller lines as needed. One or more 1½-inch (38 mm) lines may be necessary for exposure protection. The next arriving tanker or engine supplies the first engine by pumping directly into the 2½-inch (65 mm) gated pump intake. Excess water may be used to refill the first engine's tank. The water supply unit then returns for another load of water.

The second water supply unit, upon arrival, sets up a portable tank and pumps or dumps its load into the tank with a hard suction hose or waterway and also returns for another load of water.

In the meantime, the engineer of the first engine, with assistance, puts the hard suction hose into the portable tank. When the supply from the first tank is depleted, the engine takes suction from the portable tank, keeping the first engine's tank supply in reserve.

Tankers with rear or side dump valves should back in unless there is a circular drive or area in which to turn around. Two portable tanks can be set up and connected to provide a larger capacity unloading site. A large nurse tanker can be used in lieu of a portable tank if space permits. (Figure D.3).

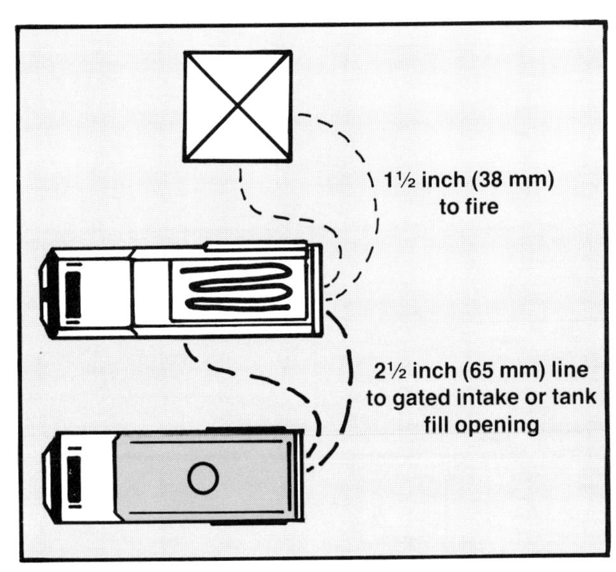

Figure D.2 Evolution 1.

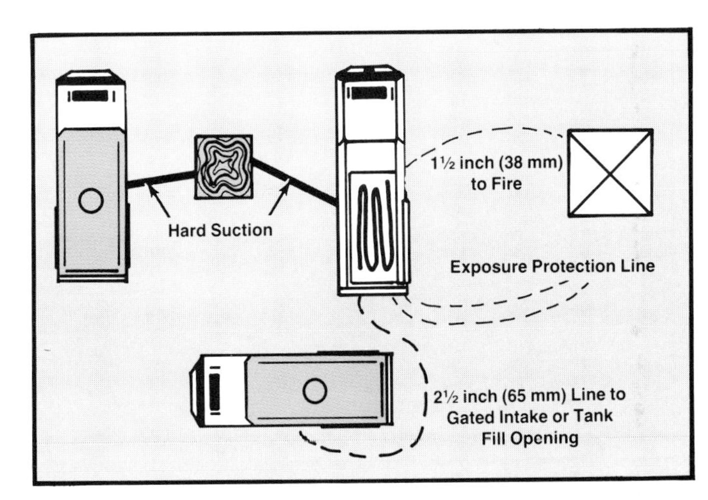

Figure D.3 Evolution 2.

Evolution 3: Large Rural Fire with Limited Farmyard

Response: 1st Alarm — Engine, Engine and Tanker or Two Tankers, Squad or Equipment Vehicle
2nd Alarm — Engine, Two Tankers
2nd Fire — Engine, Tanker, Squad or Equipment Vehicle

This evolution is an efficient way to supply an engine with tankers when there is limited room on the fireground. Tankers remain on the road or highway, thus keeping the lane and fire scene free of congestion.

The first engine lays a 2½- or 3-inch (65 mm or 77 mm) supply line from the main road to the fireground as it responds up the lane. When dropping the line at the main road, a clappered siamese must be attached.

Fire fighting operations are initiated from the tank using 1½- and/or 2½-inch (38 mm and/or 65 mm) preconnected lines needed for fire control.

As they arrive, tankers connect to alternate sides of the siamese and pump their water up the lane to the first engine. In order to maintain a steady water supply with minimum discharge problems, the second-arriving tanker waits until the first tanker has pumped its tank dry before it starts to pump. The third tanker to arrive will take the position vacated by the first tanker, and so on. A short preconnected unloading line on the tankers will speed up this operation.

The supply line can be put into a tank fill opening or gated intake of the first engine while the first tanker is unloading and the portable tank is being set up. When the portable tank is ready, the supply line can be transferred to feed the portable tank (or tanks if those from mutual aid departments are used) and the first engine can draft from the tank(s). A portable tank filling will alleviate the need for personnel to hold the supply line.

An alternate evolution would be to pump continuously into a tank fill opening (or 2½-inch [65 mm] gated pump intake of the first engine); however, using the portable tank permits a residual supply and eliminates any delay (Figure D.4).

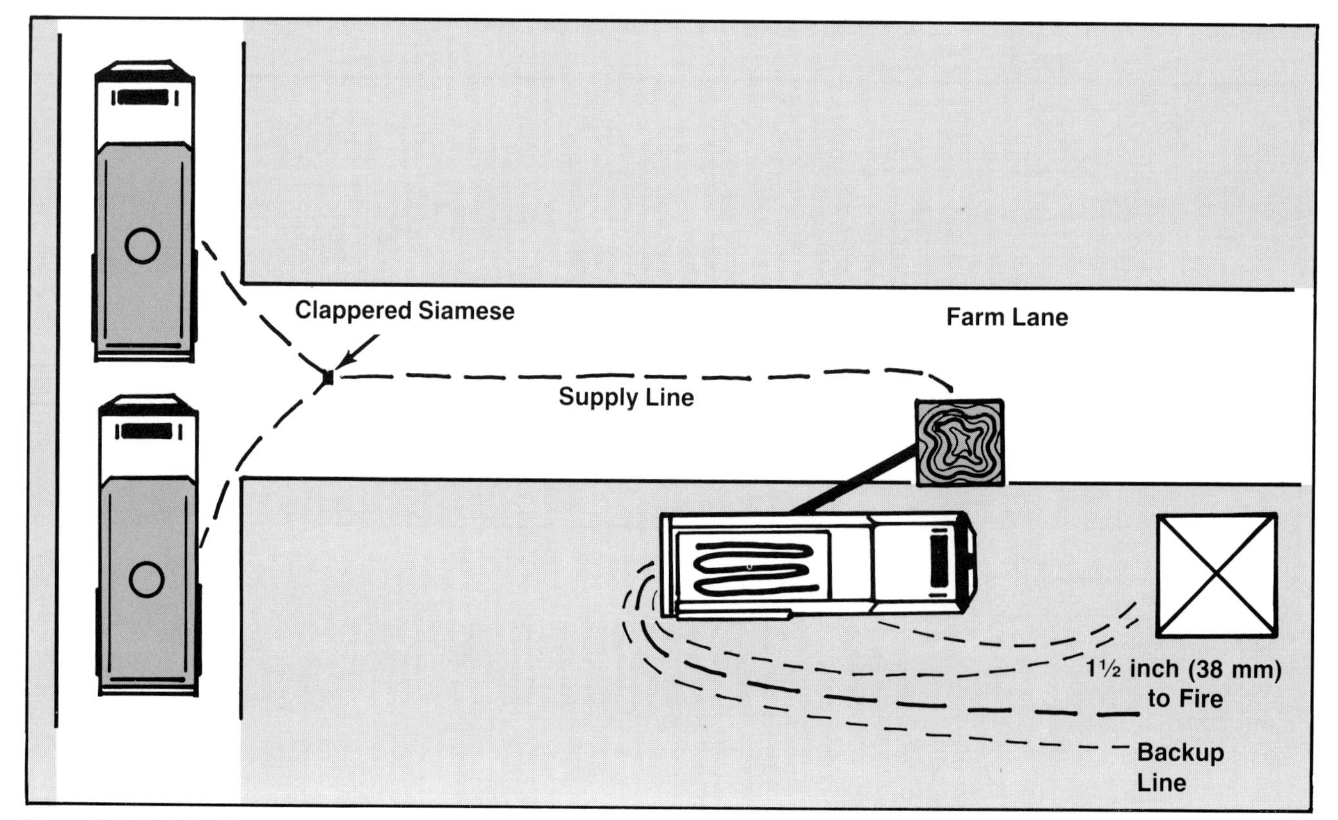

Figure D.4 Evolution 3.

Evolution 4: Large Rural Fire with Apparatus Not Equipped with Volume Pumps

Response: 1st Alarm — Engine, Engine and Tanker or Two Tankers, Squad or Equipment Vehicle
2nd Alarm — Engine, Two Tankers
2nd Fire — Engine, Tanker, Squad or Equipment Vehicle

This evolution may be used in place of Evolution 3 to maintain an effective water supply in a limited area. This operation is useful where tankers are not equipped with volume pumps and the second engine can be utilized to pump an adequate supply to the attack pumper from an unloading site.

The first engine, rather than laying in a supply line, proceeds directly to the fire and begins the initial attack. The second-arriving apparatus (engine or tanker) will back in, connect the first engine and refill the tank of the first engine. When the tank of the second unit is depleted, it will make a reverse lay down the lane to the road and pump water from subsequent arriving tankers to the attack pumper.

A nurse tanker or portable tank can be utilized at the road for a residual supply reservoir and to empty shuttling tankers. Additional water can be stored at the scene by either placing a portable tank at the end of the supply line for the attack pumper to draft from or by placing additional portable tanks at the collection point (Figure D.5).

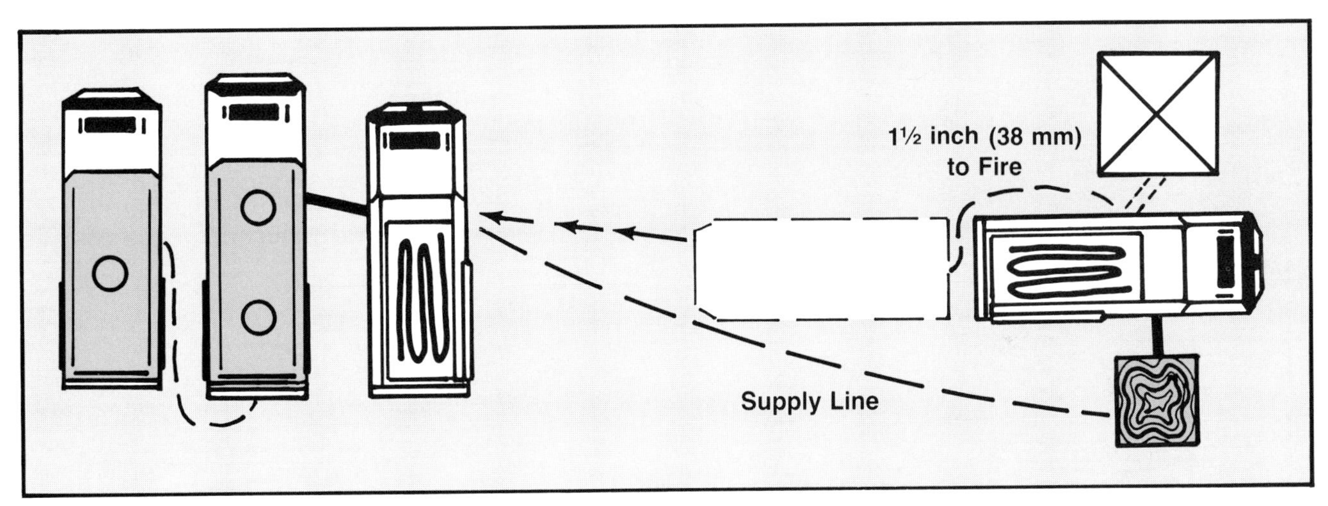

Figure D.5 Evolution 4.

Evolution 5: Large Rural Fire Utilizing a Relay Operation

Response: 1st Alarm — Engine, Engine and Tanker or Two Tankers, Squad or Equipment Vehicle
2nd Alarm — Engine, Two Tankers
2nd Fire — Engine, Tanker, Squad or Equipment Vehicle

Another alternative to use on long lanes to alleviate traffic congestion is relaying water from the main road to the fireground. This can be done by having the first engine start to lay hose at the drive, emptying its bed except for necessary fire fighting line, and then proceeding to the building for fire attack. The second engine would then back down the lane, or go in and turn around, and lay line from the attack pumper to the place where the first engine's line stopped. If the lane is over 1,500 feet (457 m), the second engine should connect into the lines here and relay from a third engine at the main road. If the lane is shorter, the second engine can connect the two lines and proceed to the main road and pump from there to the attack pumper (Figure D.6).

Another method of starting the relay would be to make a split lay. In this method, the first engine would proceed down the lane to a distance within reach of its hose bed and then begin laying hose onto the fireground. The second engine would then back down the lane to the start of the first engine's line and then make a reverse lay to the main road.

For lanes over 1,500 feet (457 m) a three-engine relay will be best. The third engine would pick up at the second engine and proceed to the main road. A two-line relay will provide a better water supply to the fireground and should be used when apparatus, hose, and an adequate water supply are available (Figure D.7).

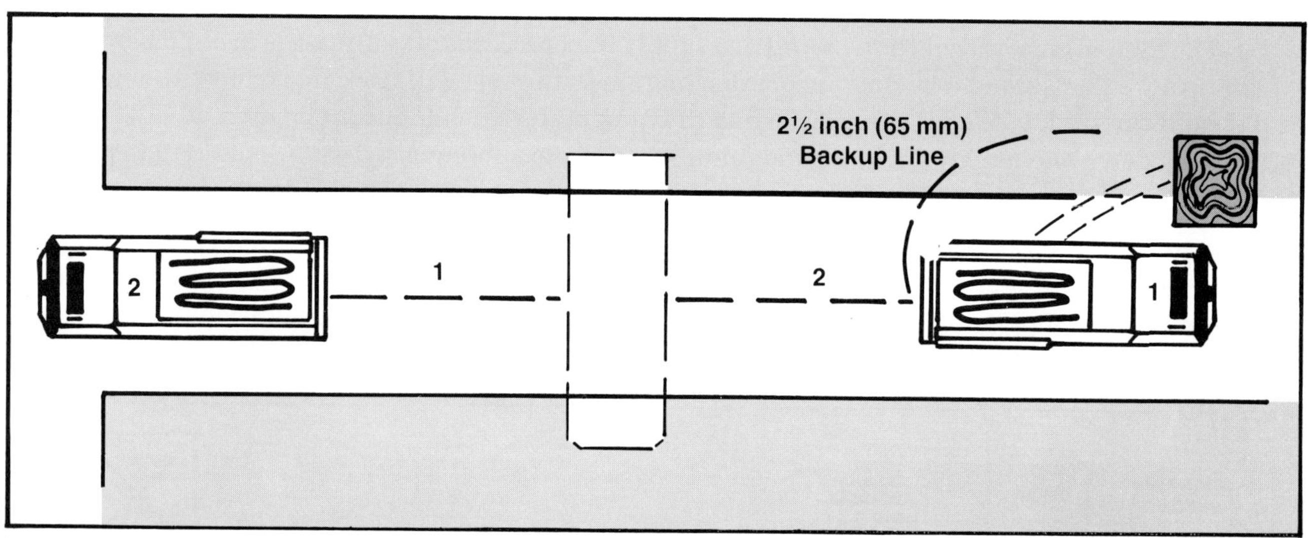

Figure D.6 Evolution 5a.

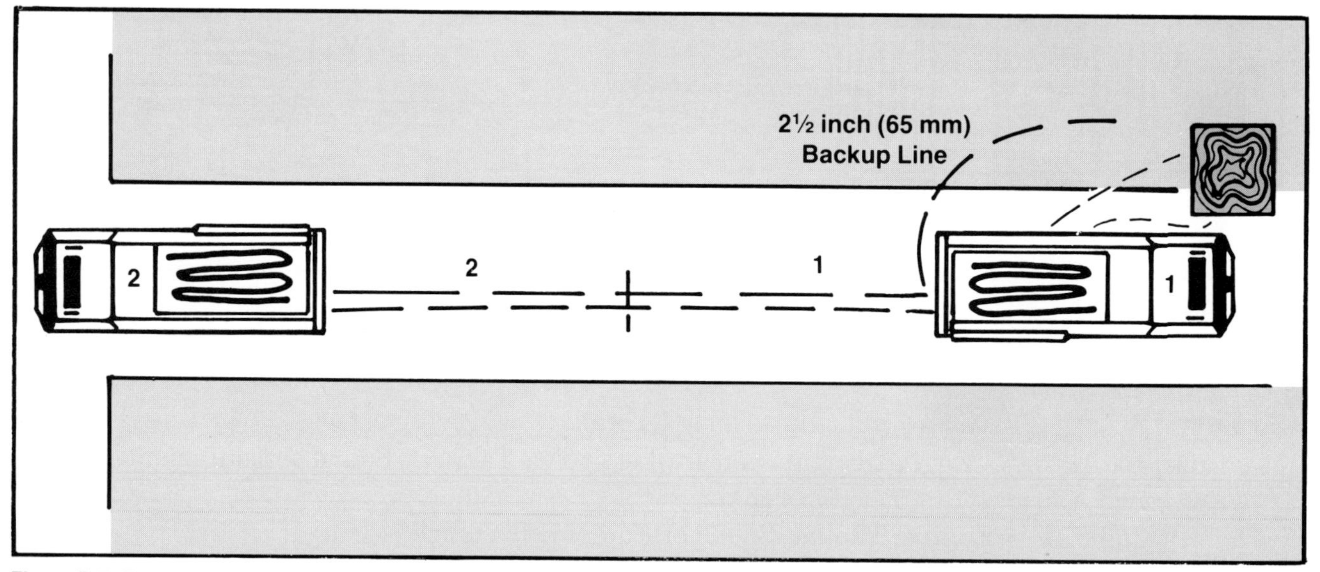

Figure D.7 Evolution 5b, showing a two-line relay.

Evolution 6: Large Rural Fire with Limited Response

Response: 1st Alarm — Engine Company, Tanker, Squad or Equipment Vehicle

When a second piece of apparatus is not available to pump water up a long lane, the first engine should lay a supply to the fireground. Then a portable pump can be set at the main road and pump the water from a portable tank or nurse tanker to the attack pumper. This allows tankers to dump and go rather than pump their load to the fireground. This operation is useful with tankers having large dump valves and no volume pump (Figure D.8).

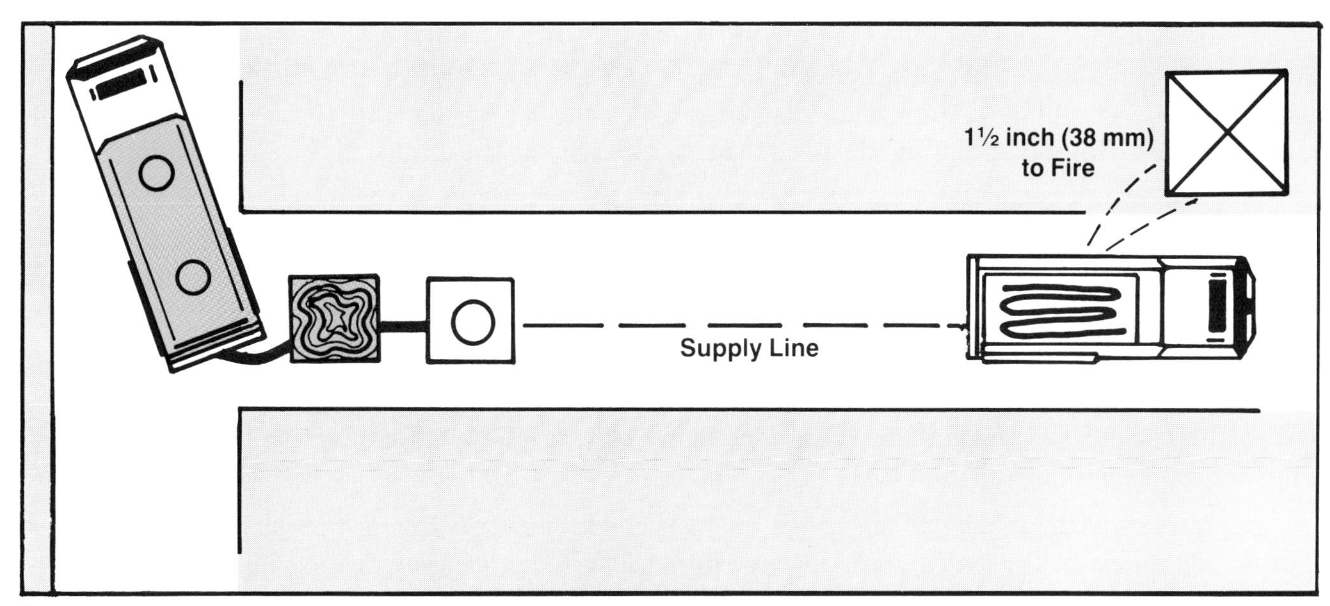

Figure D.8 Evolution 6.

Evolution 7: Filling Tank Vehicles

Several efficient means of filling tankers can be utilized. Tank vehicles with one or preferably two valved direct tank fill lines are the quickest and most reliable method for filling tanks. A large top tank opening can also be used, and in all cases proper venting is necessary.

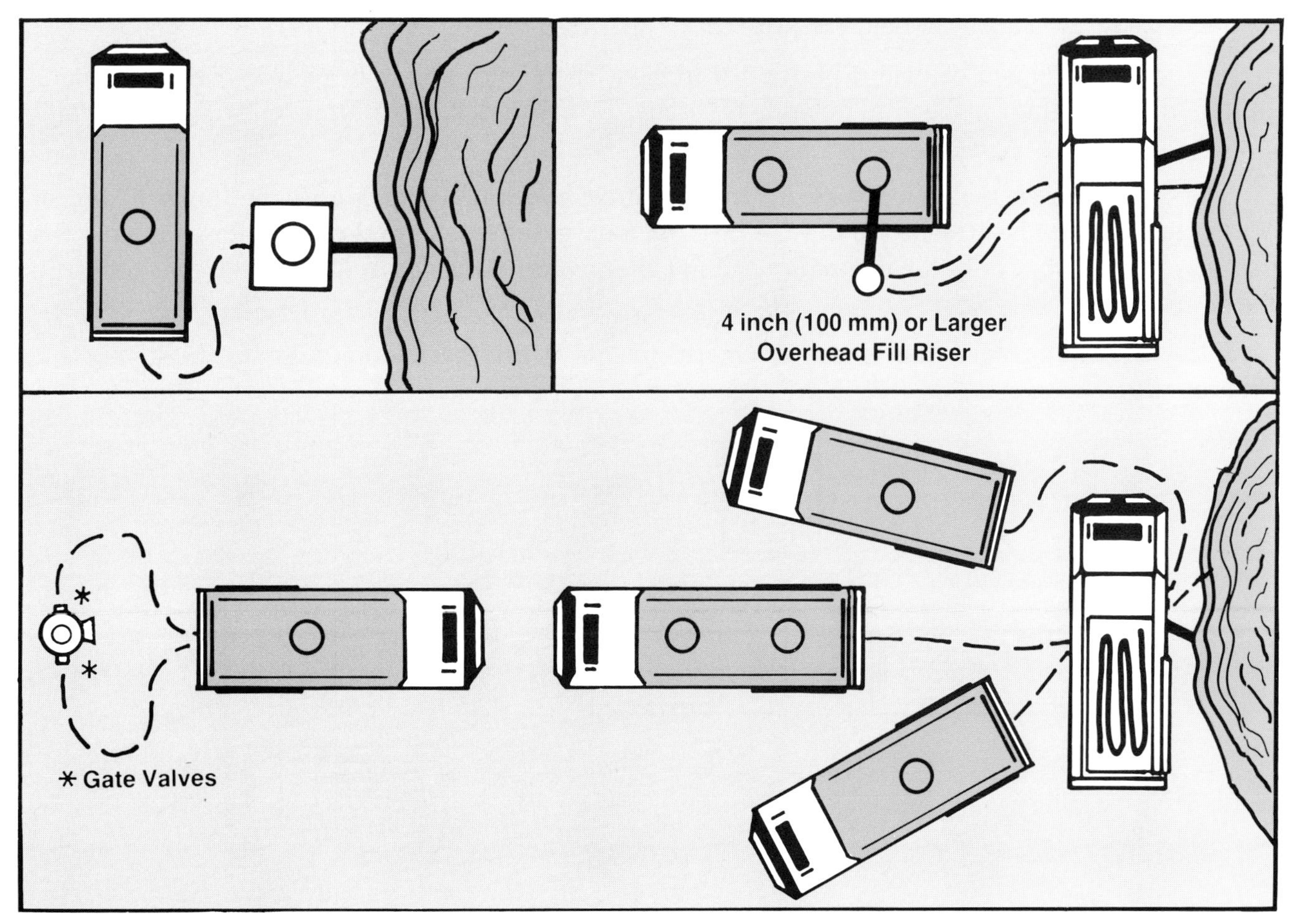

Figure D.9 Evolution 7.

A portable pump and necessary equipment can be placed at a static water supply and used to fill tankers. This operation will be limited by the capacity of the portable pump(s). Rather than stopping the pump between filling tankers, it can discharge at a reduced rate back into the source until another tanker arrives. This will eliminate the need to restart and prime the unit each time another tanker arrives.

If tankers are filled from hydrants, a short preconnected line or two lines can be made ready. These should be gated so the hydrant does not have to be shut down each time. As a tanker arrives these can be attached or placed in the top fill, gate valves opened, and the tanker quickly filled.

When a pumper is available to fill tankers from a static source or hydrant, it can use both a large sleeve and a small sleeve to the gated intake to bring maximum water to the suction side of the pump. Two or three short, but unequal, discharge lines can be taken off the fill pumper to fill several tankers simultaneously or one with dual lines.

For tankers with large top fill openings, an overhead fill device utilizing 4-inch (100 mm) or larger pipe is very effective. Two or three 2½-inch (65 mm) inlets should be provided to supply the riser (Figure D.9).

Evolution 8: Residential Fire in the Community

Response: 1st Alarm — Two engines; Ladder, Squad or Equipment Vehicle
2nd Alarm — Same as first alarm

First engine upon arrival stops just past the fire building and attacks the fire using 1½-inch (38 mm) lines as necessary, supplied from its tank. Additional lines may be needed to protect exposed buildings or the area above or below the fire.

The ladder or squad vehicle is positioned in front of the building to provide necessary ladders, equipment, and lights for the vital function of rescue, forcible entry, ventilation, opening up, salvage, and overhaul.

The second engine makes a reverse lay to the nearest hydrant laying two 2½- or 3-inch (65 mm or 77 mm) lines; one to be attached to the 2½-inch (65 mm) gated intake to the pump for supplying the first engine, the other to be a backup line and not charged until needed and ordered by the officer in charge.

If the main pump suction is gated, it is more efficient to use this suction inlet for the supply line (Figure D.10).

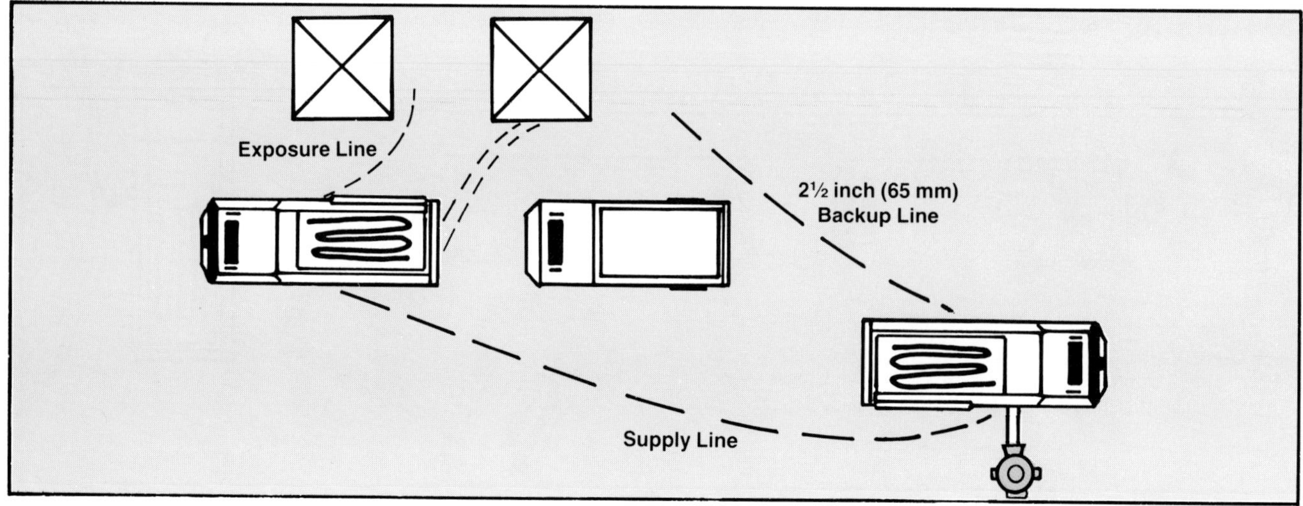

Figure D.10 Evolution 8.

Evolution 9: Residential Fire — Alternate for Heavily Involved Dwelling

Response: 1st Alarm — Two Engines; Ladder, Squad or Equipment Vehicle
2nd Alarm — As above

The first engine, upon approaching the fire, lays its own 2½- or 3-inch (65 mm or 77 mm) supply line using a gate valve at the hydrant. Depending on hose arrangement other fittings may be necessary.

The hose clamp is placed 5 feet (1.5 m) in front of the last coupling off the hose bed. The fire is initially attacked from the tank using preconnected lines while the supply line is connected into a gated pump intake.

The ladder, squad, or equipment vehicle should be positioned in front of the building so the ladders and tools can be used as needed for rescue, forcible entry, ventilation, salvage, opening up, and overhaul.

The second engine covers the rear of the building if possible, making a reverse lay to a nearby hydrant laying two lines. One can be wyed to two 1½-inch (38 mm) lines and used immediately for extinguishment and overhaul. The second line, the 2½-inch (65 mm) backup, is not charged until needed and ordered by the officer in charge (Figure D.11).

This operation could be improved by the use of a four-way hydrant valve or Humat valve, which would enable the first engine to have water immediately. Then a later-arriving engine could hook up to the hydrant through this valve and pump to the first engine (Figure D.12 on next page).

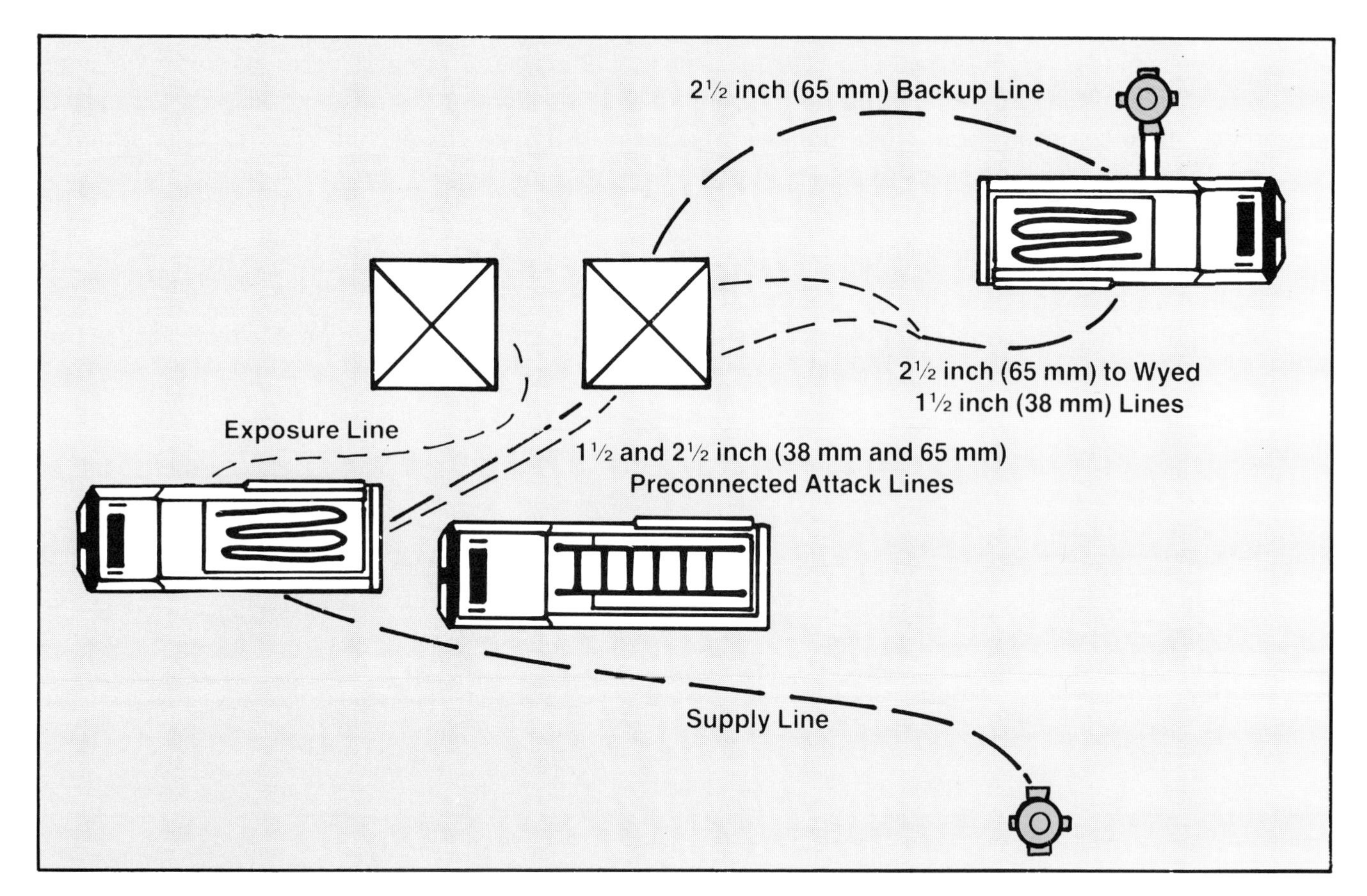

Figure D.11 Evolution 9a.

A four-way valve can be improvised by putting a short piece of hose and clappered siamese on the hydrant end of the first responding engine's supply line. This is placed on one outlet of the hydrant and

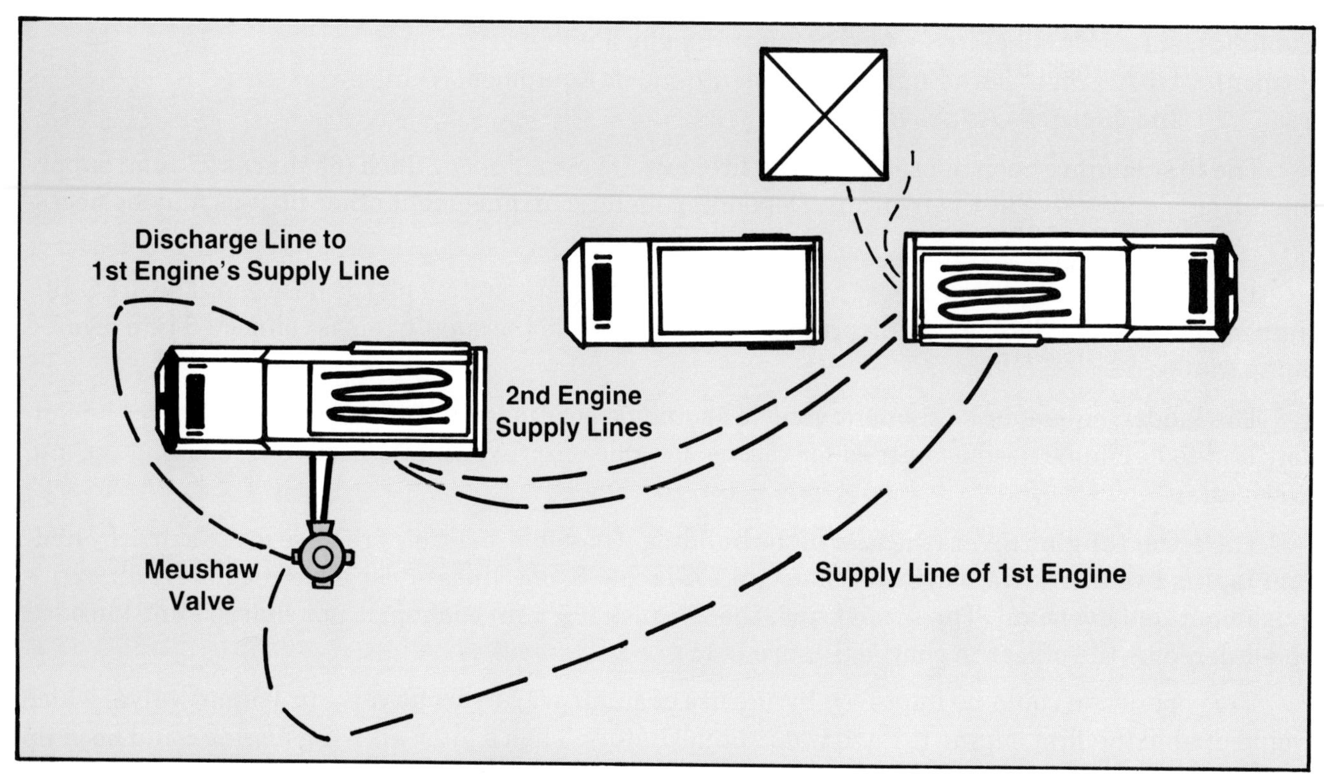

Figure D.12 Evolution 9b, showing use of four-way hydrant valve.

a gate valve is placed on the other. The hydrant is opened and the first engine receives water through the ungated side. The engine hooking up to the hydrant takes water from the gated side through a large suction to the pump and then places a discharge line into the second side of the clappered siamese. When the engine begins to pump, water will flow from the hydrant to the pump and be discharged into the supply line, closing the clapper on the direct hydrant side (Figure D.13).

The second engine, which functions as a water supply company, may back down in front of a building and make a reverse lay to the hydrant and then pump through supply line laid by first engine and any additional lines the second engine lays itself. A third engine would then be needed to cover the rear of the fire building.

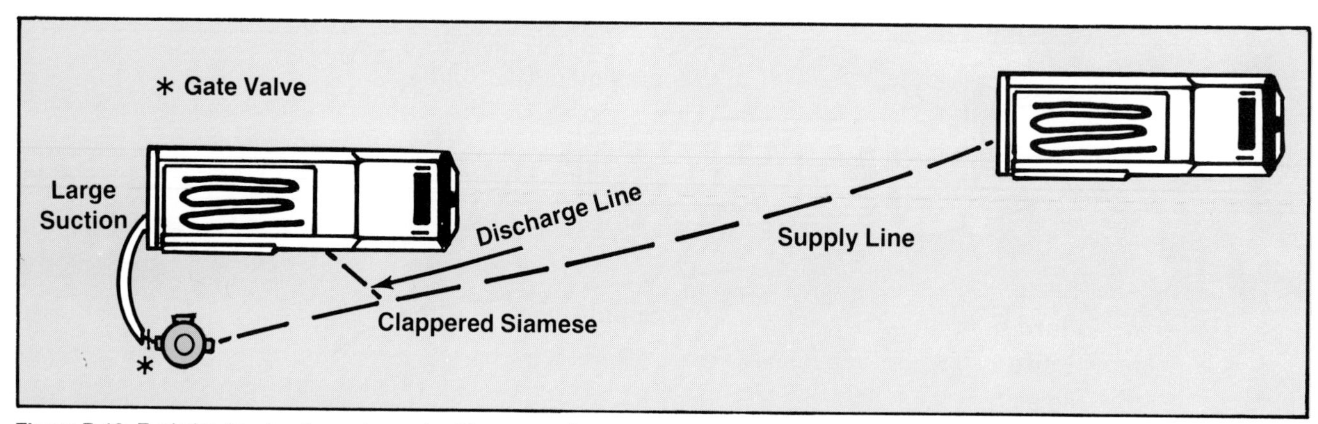

Figure D.13 Evolution 9c, showing an improvised four-way valve.

Evolution 10: Commercial, Institutional, and Industrial Buildings

Response: 1st Alarm — Two Engines; Ladder, Squad or Equipment Vehicle
2nd Alarm — Three Engines
2nd Fire — Two Engines; Ladder, Squad or Equipment Vehicle

On arrival at the fire, the fire engine drops sufficient hose for 1½- and 2½-inch (38 mm and 65 mm) attack lines, fittings, tools, ladders, and necessary equipment and makes a reverse lay to the nearest hydrant. Initially, water can be supplied from the tank while the soft sleeve is being hooked onto the hydrant.

The second engine covers the rear of the fire building. Sufficient hose for attack lines is dropped and two supply lines are laid to the nearest hydrant. One line may be wyed to 1½-inch (38 mm) if smaller lines are needed.

As conditions warrant, the ladder truck or the squad or equipment can be used for rescue, forcible entry, ventilation, salvage, opening up, and overhaul (Figure D.14).

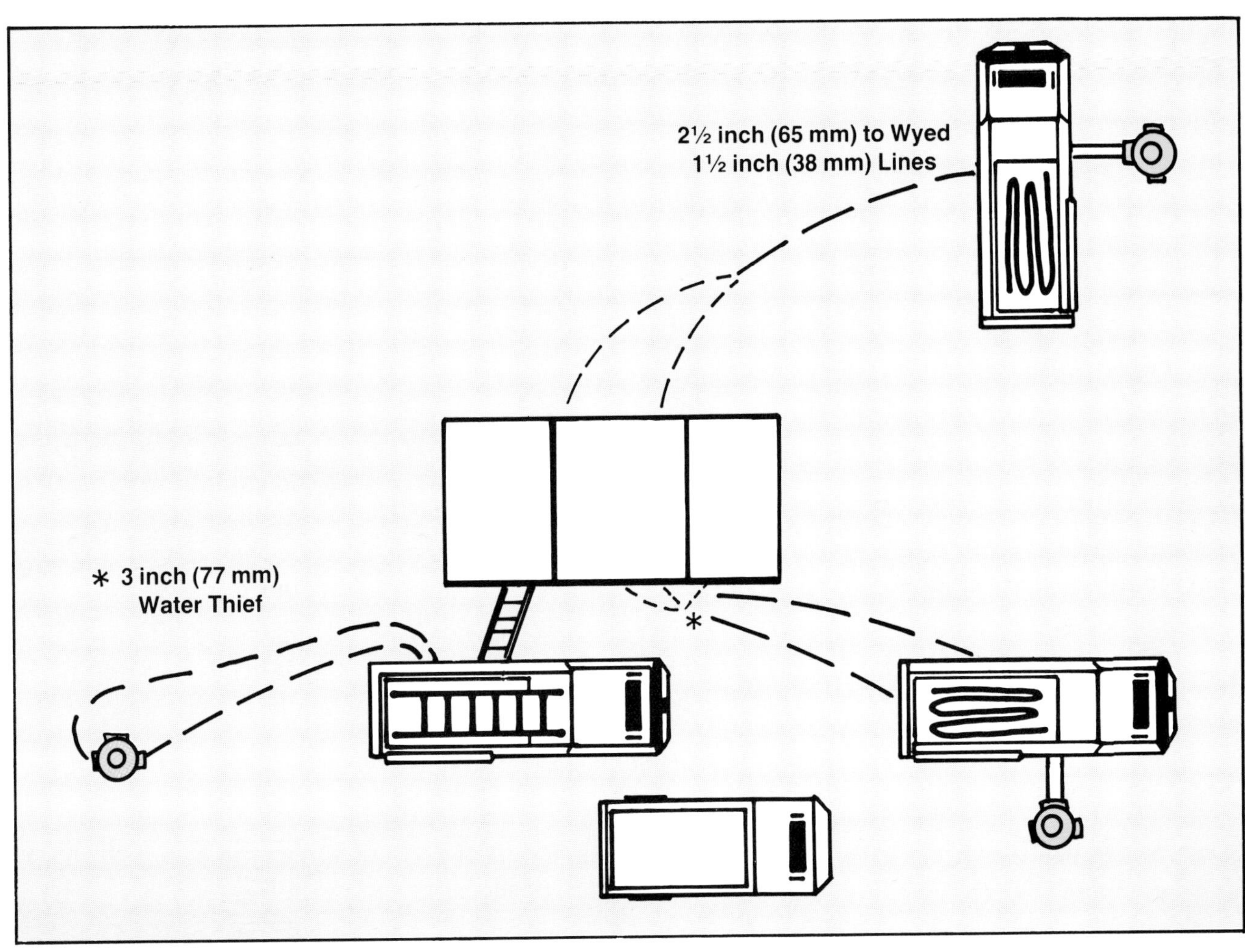

Figure D.14 Evolution 10.

Helpful Hints for Large or Mutual Aid Shuttle Operations

- In a standard shuttle, as shown, the first-due mutual aid pumper should respond to the designated fill site to handle the refilling of tankers (Figure D.15 on next page). This pumper should use at least two 2½-inch (65 mm) lines to fill tankers. With crews at both the filling and unloading sites, tanker driver/operators should *never* need to leave their cabs.
- With most midship pumps, departments can draft from both sides at the same time (Figure D.16 on next page). Closing down an empty portable tank can be done with a gate valve, suction cap from the pump intake, or a volleyball. (**NOTE:** The last two methods will not work when using a strainer.)

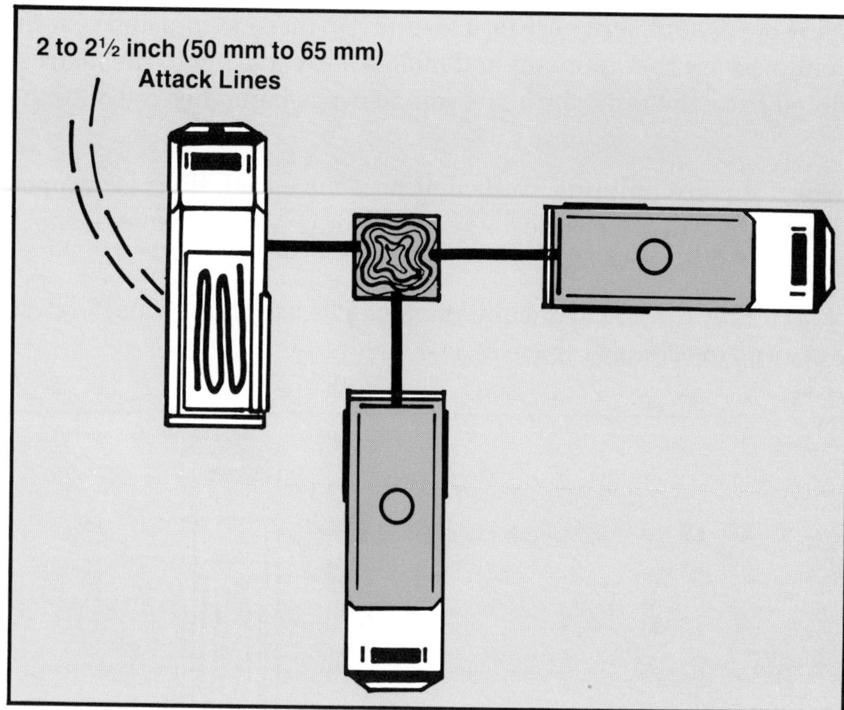

Figure D.15 A standard shuttle. If crews are at both sites, tanker driver/operators should not have to leave their cabs.

Figure D.16 Most midship pumps can draft from both sides at the same time.

- Setting portable tanks as shown in Figure D.17 gives tankers better access to dump loads. As an added bonus, operators have more room to work around pump panels and tanker dump valves.

- Two pumpers working a relay fire attack can pump distances of up to ½ mile (0.8 km). Beyond that it is more efficient to use a tanker shuttle.

- When relay pumping, pump vacuum gauges should read at least 20 psi (138 kPa) on pumpers receiving water from another pumper.

- By using a gated wye, pumpers can switch portable tanks when one becomes empty (Figure D.18).

- When portable tanks are full and a tanker has to wait at the unloading site to dump its load, either more portable tanks or fewer tankers are needed.

- If it is possible for only one tanker to unload at a time, the portable tank may be poorly positioned. Under ideal conditions, up to three tankers should be able to unload at the same time.

- In situations where there is limited access to portable tanks, tankers equipped with high volume pumps (500 gpm [1 893 L/min] or greater) may pump off their water through hoselines to the portable tanks.

- Higher road speeds will not provide greater water flows. High volume loading and unloading of tankers is the key to greater water flow capabilities.

- Drafting devices, such as floating dock strainers and portable tank strainers, will ensure maximum use of the total water available in the portable tank.

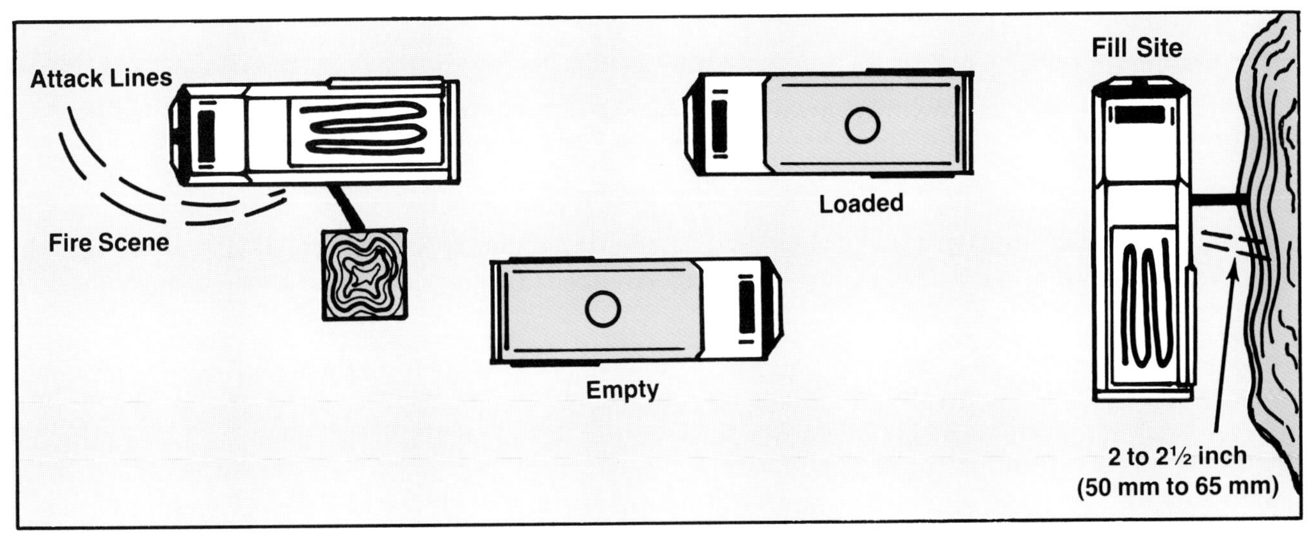

Figure D.17 Setting portable tanks as shown gives tankers better access and gives personnel more room to work.

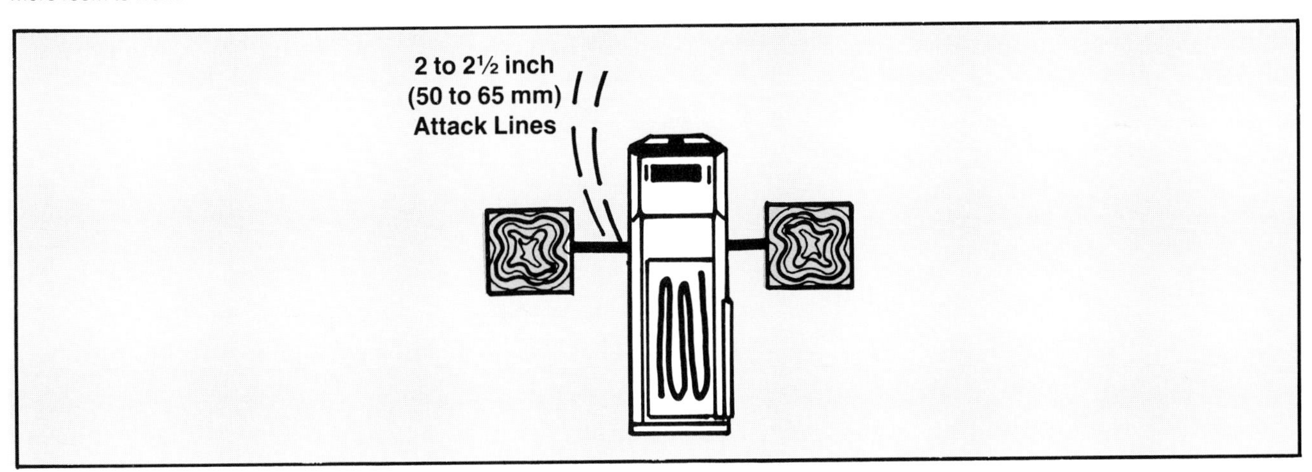

Figure D.18 A gated wye enables pumpers to switch tanks when one becomes empty.

POSITIONING OF ENGINE COMPANIES FOR CITY ALARMS

When Using a Straight or Forward Lay

1. The first engine should be positioned just beyond the fire building when laying a hydrant supply line or working from its tank. The initial incident commander is then able to size up three sides of the building upon arrival. This is also the best pumper position for utilizing rear preconnected attack lines. The truck or squad company can be positioned directly in front of the building for efficient use of the equipment.

2. When the first engine has laid a supply line or is working from its tank, the second-arriving engine company will take one of the following positions:

   a. Stop at the hydrant and pump to the first engine through the supply line laid by the first engine.

   b. Back down and reverse lay one or two additional supply lines to the hydrant from which the first engine laid its line.

   c. If the first engine did not lay in, the second-arriving engine should automatically reverse out two supply lines.

   d. Cover the rear of the fire building if so ordered by the incident commander.

3. Generally, a third engine company will be necessary to cover the rear of the building.

When Using a Reverse Lay

1. The first engine stops just beyond the fire building prior to making a reverse lay to a water source. This aids in size-up and provides room for pulling skid or necessary hose for attacking the fire. All equipment needed on the fireground must be removed from the engine before it goes to the hydrant.
2. The second engine takes one of the following positions:
   a. Backs down the alley or goes to the rear of the building and makes a reverse lay to the nearest hydrant to cover the rear of the fire building. All needed equipment must be removed from the apparatus before going to the hydrant.
   b. Backs down to the front of the building and makes a reverse lay from the first engine to provide additional water supply in the front of the building.

When Using a Two-Piece Engine

1. When using a two-piece engine acting as a hose wagon and a pumper, the hose wagon takes a position just beyond the fire building, and the pumper connects to the hydrant four-way valve or picks up the dry line laid by the wagon and connects to the hydrant.
2. With two-piece engines, the second company should cover the rear and set up using the normal standard operating procedures.

PRECAUTIONS IN POSITIONING

- Do not get too close to the fire building or too far from it.
- Try to get a position upwind.
- Do not block incoming apparatus.
- Do not block the removal of ladders from ladder trucks or other equipment from ladder or squad trucks by parking too close.
- Do not take a position under electric power lines.
- Do not get on soft ground, lawns, or sand where apparatus will get stuck and be out of service.
- Do not block good hydrants that can support additional pumpers.
- When working as an attack engine with a water supply engine, the second engine should not go directly into the fire area and block the first engine. Rather, the second engine should stop at the last hydrant and wait for orders to bring a line in or back down and prepare to reverse lay to a hydrant.

Appendix E
Flow Rating Tanker Apparatus

By William F. Eckman
LaPlata, MD

Any water shuttle operation consists of three parts: the tanker must first dump its load, then travel to a water supply, and finally fill up again. Then, of course, the tanker has to return to the scene of the emergency.

To determine how much water a given tanker can supply, it is necessary to determine the time required for a complete round trip, including dumping and filling. Then divide this time into the capacity of the tanker to arrive at gallons per minute (L/min) figure. This calculation should be available for each tanker that is likely to respond to an emergency. These figures can also be used in determining how many units will be required for a particular fireground situation. Figures E.1-E.3 show record and evaluation sheets that may be used to record test data and assist in calculations.

TANKER RATING

By measuring the fill rate and dump rate, it is possible to calculate the amount of water that a specific water supply apparatus can supply over various distances. For purposes of estimating supply capabilities, the time required for each trip can be divided into two categories, Travel Time (TT) and Handling Time (HT). The formula used for determining travel time by the Insurance Services Office (ISO) in rating Fire Department Water Supply is:

TT in minutes (U.S) = .65 + (1.7) (Distance in miles)
TT in minutes (Metric) = .65 + (1.06) (Distance in km)

This formula allows approximately 40 seconds for acceleration as the vehicle leaves the fill site and dump site and for deceleration as it approaches these sites. It assumes an average speed of 35 mph (56 km/h), which must be adjusted for specific situations. If road conditions are good, with few or slight grades, a somewhat higher average speed may be possible. For example, an emergency on a four-lane limited access highway may involve hauling water for some distances at relatively high speeds. If road or weather conditions are hazardous, however, speeds will also increase the travel time, but overall, this formula provides a reasonably accurate estimate of the travel time actually required.

Handling time includes the following evolutions:

Filling the Tank (Time actually flowing water)
Dumping the Tank (Time actually flowing water)
Connecting and Breaking Hoselines
Maneuvering on the Fill Site and Dump Site

The fill rate of a given tanker is primarily determined by the construction of the tank. Important factors include type of piping, venting, and construction of the baffles within the tank. The flow rate is limited by the size of the inlet piping and the way it is connected. Each fitting, such as an elbow or a tee, increases friction loss, reduces the maximum flow rate, and requires a higher pressure to fill the tank.

While the tank is filling with water, the air inside the empty tank is forced out. If the vent on the tank is inadequate, there will be a pressure buildup inside the tank that reduces the fill rate and may result in damage to the tank. The internal baffling may restrict the movement of water from the fill point through the rest of the tank; it may also restrict the movement of air toward the vent. If this hap-

WATER SHUTTLE RECORD SHEET

Date ____________________ Location ____

Unit # ____ Capacity ____________________

RECORD OF OPERATION

	Trip #1	Trip #2	Trip #3	Average
Arrival at Dump Site				
Begin to Dump				
Tank Empty				
Leave Dump Site				
Dump Time				
Arrive at Fill Site				
Travel Time				
Begin Filling				
Tank Overflows				
Leave Fill Site				
Fill Time				
Arrive at Dump Site				
Travel Time				
Total Time — Round Trip				
Waiting Time: Fill Site				
Dump Site				
Handling Time				
Travel Time				

EVALUATION

Capacity ____________________ Number of Trips ________________ Total Gallons ____________

Avge. Handling Time ____________ Avge. Travel Time ____________ Total Time/Trip ____________

GPM actually supplied ____________________ Corrected to GPM/Mile ____________________

Figure E.1 A water shuttle record sheet.

EVALUATION BY WEIGHT

	Frt Axle	Rear Axle	Total	Ratio
W1 — On Arrival				
W2 — After maximum discharge				
W3 — Completely drained				
W4 — Completely filled				

	Weight	Gallons
Full Load (W4 — W3)		
Spillage (W4 — W1)		
Usable Water (W1 — W2)		
Ballast (W2 — W3)		

RESULTS OF WATER SHUTTLE EXERCISE

One way distance ____________________

Average travel time ____________________

Average fill time ____________________

Average fill rate ____________________

Average dump time ____________________

Average dump rate ____________________

Average handling time ____________________

Actual GPM supplied ____________________

Actual GPM/mile ____________________

Corrected GPM/mile ____________________

Figure E.2 Evaluation by weight and results of water shuttle exercise can be recorded on forms such as these.

APPARATUS EVALUATION SHEET

UNIT #__________ FIRE DEPARTMENT __________

Tested at __________ By __________ Date __________

SPECIFICATIONS

Chassis: Mfr. __________ Model __________ Year __________

Engine — Type __________ Model __________ CU Disp. __________

Weight Rating — GVW __________ FA __________ RA __________

Body: Type __________ Mfr. __________

Tank: Shape __________ Capacity __________

Pump: Type __________ Mfr. __________ Capacity __________

TEST RESULTS

DISCHARGE TESTS — Pumping				FILL TESTS	
Nozzle Pressure	____	Size Dump	____	Source	____
Size tip or flow rating	____	Jet Pressure (Jet Dump Only)	____	Fill Lines	____
Estimated Flow	____	Discharge Time	____	Fill Time	____
Discharge Time	____	Dump Rate	____	Fill Rate	____
Usable Water (Calculated)	____	Usable Water	____	Capacity	____

Calculated GPM/Mile: Dumping __________ Pumping __________

EVALUATION — PRACTICAL EXERCISE

Total water hauled __________
Capacity — usable water __________
Average fill time __________
Average fill rate __________
Average dump time __________
Average dump rate __________
Distance traveled __________
Average travel time __________
Calculated GPM __________

Figure E.3 An apparatus evaluation sheet.

pens, the vent may overflow while portions of the tank are not full of water. This is a greater problem when the tank fill line discharges at the same end of the tank as the vent location. Air pockets may be trapped in one end of the tank and water may come out the overflow before the tank fills.

A major consideration in the fill time required on a particular water shuttle operation is the capacity of the supply. If the tankers are being filled from a hydrant system, the maximum fill rate is limited to the maximum flow of the particular hydrant being used. If the tankers are being filled from draft, the maximum fill rate is limited by the capacity of the supply pumper, the size and number of hoselines being used, and the amount of water available in the source.

The time required to dump the tanker is also dependent upon the construction of the tanker. Again, the piping, the venting, and the baffling are important. Many fire apparatus are equipped with pumps that are larger than the tank-to-pump piping can supply. If the primary means of discharge is by the use of the pump on the tanker, it should be supplied by the largest piping possible. Many tankers have doubled the dump rate by changing the size of the tank-to-pump line. The elimination of a few elbows or tee fittings can increase the flow rate by as much as 20 percent, even with the same size line.

For the maximum flow rate, the pump should be operated on the verge of cavitation. Cavitation occurs when an increase in throttle setting and a corresponding increase on the intake gauge indicate a vacuum reading of more than 20 inches (510 mm). If the tank is inadequately vented, the atmospheric pressure may be reduced by an inadequate air flow to replace the water that has been removed. This reduces the pressure differential that is available to force the water from the tank to the pump, and the pump may drift into cavitation. If cavitation occurs, the rpm must be decreased with a corresponding reduction in the flow rate and increase in dump time. Inadequate baffling can create a similar condition if the transfer rate between sections of the tank is less than the discharge rate. When this happens, the pump may run out of water while there is still water in other portions of the tank.

When a gravity dumping arrangement is used, the rate of flow will decrease as the level of the water in the tank and the resulting head pressure is reduced. This reduction is at an exponential rate, and it will require an inordinate amount of time for the last few gallons (liters) to be discharged from the tank. It is a matter of judgment as to just when to stop the discharge and send the unit for another load of water. If the discharge is shut off too soon, an excessive amount of water will remain in the tank and be hauled back and forth as ballast. On the other hand, if too much time is spent waiting for the last drops of water to drain, the average flow rate will be decreased and the gpm (L/min) delivered will be reduced. The use of a jet dump will eliminate this problem. By using the pump pressure to increase the flow through the large dump line, the rate of flow will remain constant during the entire discharge time; this will greatly increase the average discharge rate. If a tanker is equipped with a pump, even a small PTO or portable unit, a jet assist can be easily added to gravity dump, greatly increasing the capability of the unit. Some departments have constructed a portable eductor that can be externally connected and supplied for those units not equipped with a pump.

If a tanker is equipped with a large dump fitting with standard fire department threads, the dump rate can be increased by drafting from the tank with a Class A pumper. While the actual flow rate is increased, the additional time required to make connections and maneuver on the dump site increases the time required to empty the tank instead of decreasing it.

Another factor in the time required to dump a tanker is the need for adequate storage on the dump site. For the tanker to discharge its load at the maximum rate, adequate storage is necessary. Either by use of nurse tankers or portable reservoirs, sufficient capacity must be available to allow the units in the shuttle to off-load immediately on arrival, then return to the fill site for another load. Experience has shown that a minimum of three times the needed fire flow must be available in storage at the dump site in order to maintain a consistent supply, and five or six times the maximum rate is desirable.

The other two factors in handling time — making and breaking connections and maneuvering apparatus — provide the keys to a successful water shuttle operation. Fire departments driver/operators often operate tankers at a high rate of speed, endangering civilian traffic as well as fire department personnel, to save a few seconds. Then they lose several minutes in making and breaking hose connections or in maneuvering at either the fill site or the dump site. Organization is the only way to minimize loss of time here. There must be a dedicated crew of hose handlers who determine the best way to load and unload each unit in the minimum amount of time and know which hoselines to use. Traffic control is important and each tanker should be directed to the location where it is needed with a minimum of confusion.

WATER SUPPLY CAPABILITY

Appendix Tables E.1, E.2, and E.3 detail the capabilities of different size tankers with varying flow rates over some typical distances involving water shuttles. In making these calculations, estimations were made using the same standards used by the Insurance Services Organization in rating a fire department. The handling time for the dump site and the fill site is tested by having the tanker take a position approximately 200 feet (60 m) from the site. The tanker is then timed from that point to the site, filled with water, then returned to the starting point. Handling time includes the time required to get into position, to make and break the needed connections, the actual transfer of water from the source into the tank, and from the tank into the reservoir on the dump site. For purposes of this estimation, two minutes were given for maneuvering and connecting on each end of the shuttle. These four minutes were then added to the time required for water transfer at different flow rates to provide the total handling time. The travel time was calculated by use of the formula developed by ISO and doubling the distance to the source to allow for the round trip mileage.

As an example, Table E.2 covers a 2,000 gallon (7 570 L) tank. Assuming a flow rate of 500 gpm (1 893 L/min), both fill and dump, and a distance of 1 mile (1.6 km), its capacity would be:

U.S.:

Handling Time = (Fill Site Connecting and Maneuvering) + (Unload Site Connecting and Maneuvering) + (Loading Time) + (Unloading Time)

Handling Time = (2 min.) + (2 min.) + (2,000 ÷ 500) + (2,000 ÷ 500)
Handling Time = 12 min.

Travel Time = (2 miles x 1.7) + .65 = 4.05 or ≈ 4 min.

$$\text{GPM} = \frac{2{,}000 \text{ gpm}}{12 \text{ min.} + 4. \text{ min.}} = 125 \text{ gpm}$$

Metric:

Handling Time = (2 min.) + (2 min.) + (7 570 ÷ 1 893) + (7 570 ÷ 1 893)
Handling Time = 12 min.

Travel Time = (3.2 km x 1.06) + .65 = 4 min.

$$\text{L/MIN} = \frac{7\ 570 \text{ L/min}}{12 \text{ min.} + 4 \text{ min.}} = 473 \text{ L/min}$$

Appendix Table E.1
TANKER FLOW RATES (U.S.)

Travel Distance Each Way In Miles	Handling Time in Minutes						
	6	7	8	9	10	11	12
1	100	91	83	77	71	67	62
2	77	71	67	62	59	56	53
3	59	56	53	50	48	45	43
4	50	48	45	43	42	40	39
5	42	40	39	37	36	34	33
10	24	24	23	23	22	22	21

TANK CAPACITY = 1,000 GALLONS

Appendix Table E.1
TANKER FLOW RATES (METRIC)

Travel Distance Each Way In km	Handling Time In Minutes						
	6	7	8	9	10	11	12
1.6	379	344	314	291	269	253	235
3.2	291	269	253	235	223	212	201
4.8	223	212	201	189	182	170	163
6.4	189	182	170	163	159	151	148
8.0	159	151	148	140	136	129	125
16.0	91	91	87	87	83	83	79

TANK CAPACITY = 3 785 LITERS

Appendix Table E.2
TANKER FLOW RATES (U.S.)

Travel Distance Each Way In Miles	Handling Time in Minutes						
	8	10	12	14	16	18	20
1	167	143	125	111	100	91	83
2	133	118	105	95	87	80	74
3	105	95	87	80	74	69	65
4	91	83	77	71	67	63	59
5	77	71	67	63	59	56	53
10	47	44	43	41	39	38	36

TANK CAPACITY = 2,000 GALLONS

Appendix Table E.2
TANKER FLOW RATES (METRIC)

Travel Distance Each Way In km	Handling Time In Minutes						
	8	10	12	14	16	18	20
1.6	632	541	473	420	379	344	314
3.2	503	447	397	360	329	303	280
4.8	397	360	329	303	280	261	246
6.4	344	314	291	269	254	238	223
8.0	291	269	254	238	223	212	201
16.0	178	167	163	155	148	144	136

TANK CAPACITY = 7 570 LITERS

Using the same example, increasing the flow rate from 500 gpm (1 893 L/min) to 1,000 gpm (3 785 L/min) would reduce the total handling time from 12 minutes to 8 minutes. This would allow the same tanker to supply 167 gpm (632 L/min), or one third more water, in the same situation.

From this comparison, it can be seen that longer hauls require larger tankers. In the first example, over a distance of 1 mile (1.6 km), the 5,000 gallon (18 930 L) tanker only supplies slightly more than twice as much water as the 1,000 gallon (3 785 L) unit. When the distance is increased to 10 miles (16 km), however, the larger tanker provides almost four times as much water.

The flow rate also assumes added importance as the size of the tank becomes larger. Over a 1-mile (1.6 km) distance, the capacity of the 1,000 gallon (3 785 L) tanker increased less than 20 percent when the flow rate doubled, while the capacity of the 5,000 gallon (18 930 L) unit went up by 50 percent. Over the longer distances, the increased flow rate becomes less significant as travel time becomes longer.

In designing a water shuttle, then, tankers with high flow rates are most desirable for short hauls while larger tanks will do a better job over longer distances. This should be taken into account, both in

Appendix Table E.3
TANKER FLOW RATES (U.S.)

Travel Distance Each Way In Miles	Handling Time in Minutes 14	16	18	20	24	26	30
1	278	250	227	208	179	167	147
2	238	217	200	185	161	152	135
3	200	185	172	161	143	135	122
4	179	167	156	147	132	125	114
5	156	147	139	132	119	114	104
10	102	98	94	91	85	82	77

TANK CAPACITY = 5,000 GALLONS

Appendix Table E.3
TANKER FLOW RATES (METRIC)

Travel Distance Each Way In km	Handling Time in Minutes 14	16	18	20	24	26	30
1.6	1 052	946	859	787	677	632	556
3.2	901	821	757	700	609	575	511
4.8	757	700	651	609	541	511	462
6.4	677	632	590	556	500	473	431
8.0	590	556	526	500	450	431	394
16.0	386	371	356	344	322	310	291

TANK CAPACITY = 18 930 LITERS

purchasing mobile water supply apparatus, and in calling for mutual aid units when establishing a water supply shuttle.

WAITING TIME

Many water shuttles bog down, either on the dump or fill end, when tankers get backed up and have to wait before filling or dumping. The only time a tanker is actually hauling water is when the wheels are moving over the road; anytime a unit spends waiting subtracts from the amount of water that it can supply. The wait may be caused by mechanical failure or other unforeseeable circumstances, but most often it is the result of poor planning or bad operating practices.

When waiting time is a factor, this must be added to the formula for determining water shuttle capacity:

$$\text{GPM} = \frac{\text{CAPACITY}}{\text{TT} + \text{HT} + \text{WT}}$$

If we use the previous examples and add to them a five-minute waiting time, tankers would be capable of supplying the following flows:

$$\text{GPM} = \frac{2{,}000 \text{ gal.}}{12 \text{ min.} + 4 \text{ min.} + 5 \text{ min.}} = 95 \text{ gpm}$$

OR

$$\text{L/MIN} = \frac{7\,570 \text{ L/min}}{12 \text{ min.} + 4 \text{ min.} + 5 \text{ min.}} = 360 \text{ L/min}$$

Review Answers

CHAPTER 1

1. False. The water utility's primary concern is to provide water for human consumption.
2. False. The purpose of the Fire Protection Master Plan is to address the development and direction of the community's fire protection system.
3. False. There are many other potential supply resources, including tank truck owner/operators, heavy equipment agencies and/or contractors, owners of private water sources, military installations, Civil Defense or Emergency Preparedness/Management Agencies, news media organizations, forest service officials, and Coast Guard and/or private tug boat fleet operators.
4. True
5. False. The fill site control officer is responsible for traffic control around the fill site.
6. C
7. C
8. B
9. D
10. A
11. B
12. A
13. A. *Uniform Fire Code*
    B. ISO *Guide for Determination of Fire Flow*
    C. ISO *Fire Suppression Rating Schedule*
14. NFPA 1231, *Water Supplies For Suburban and Rural Firefighting*
15. All of the following are correct:
    Additional sources, treatment, and pumping
    Looping and gridding or distribution system
    Water system extensions and expansions
    Additional storage facilities
    System modernization
    Sprinkler system incentives
    Coordination of zoning and water main planning
    Future funding
16. All of the following are correct:
    Seek advice from other experts when resolving problems.
    Offer reciprocal advice and support in areas beyond immediate water supply concerns.
    Appoint interagency liaison personnel to deal with individual contacts.
    Maintain open and honest communications.
17. All of the following are correct:
    Change in occupancy
    Change in building contents
    Additional exposures
    Water system repairs
    Weather conditions that deter the use of the pre-incident plan
18. All of the following are correct:
    Use of proper adapters and fittings
    Laying or handling of hose and couplings
    Personnel in full protective gear
    Proper lifting and pulling techniques
    Positioning of apparatus and portable tanks
    Hoselines under pressure
    Hose appliances
    Driving of apparatus
    Working on or about apparatus
    Mounting and dismounting apparatus
    Operation of top mount vents and fills
19. All of the following are correct:
    Variations in standard operating procedures
    Preparation of equipment differences
    Jurisdictional command procedures
    Communications
    Primary water source locations
    Intercompany familiarization

CHAPTER 2

1. True
2. False. The centrifugal pump is the most common type of pump used today.
3. False. Domestic water usage fluctuates more with the changing seasons.
4. False. The three basic types of pipe system layouts are the grid system, the tree system, and the circle or belt system.
5. True
6. False. It may require closing several valves to control a break.
7. False. The pressure exerted on the bottom of a container by a liquid is independent of the shape of the container.
8. C
9. B
10. D
11. D
12. Higher
13. Force per unit area
14. 62.5 pounds (28 kg)
15. 0.434 psi or 3 kPa
16. Water is drawn into the pump by gravity or creation of a vacuum. The water enters the eye of the impeller and turns 90 degrees as it enters vanes. It is thrown outward toward the inner walls of the casing. The water passes between the rim of the impeller and the casing and is ejected under pressure from the discharge port.
17. A. Average Daily Consumption
    B. Maximum Daily Consumption
    C. Peak Hourly Consumption
18. A. Gravity
    B. Direct Pumping
    C. Combination Systems
19. All of the following are correct:
    Pipes
    Valves
    Hydrants
    Meters
    Other appliances for conveying water
20. All of the following are correct:
    To maintain a ready supply of portable water
    To maintain adequate capacity for normal and emergency use without interruption of service
    To maintain reasonably uniform pressure in the water system during high demand periods
    To allow supply pumps sufficient cycle intervals, thereby reducing wear and operating costs
    To store water for fire fighting
    To provide adequate service for weak areas in the system
21. All of the following are correct:
    Inadequate size
    Deterioration due to age
    Corroded lines
    Poor maintenance
22. All of the following are correct:
    Cast iron
    Ductile iron
    Steel
    Cement
    Asbestos
    Polyvinyl chloride (PVC)
    Other plastics and synthetics
23. After each use and periodically

CHAPTER 3

1. False. Dry-barrel hydrants are most commonly used in colder climates.
2. True
3. False. Close the hydrant slowly to avoid water hammer.
4. False. They should be kept for five years.
5. False. Pipe wrenches may damage the operating nut.
6. D
7. B
8. A
9. B
   D
   A
   C

10. A. Wet-barrel. Wet-barrel hydrants are filled with water at all times.
    B. Dry-barrel. The dry-barrel hydrant contains no water when not in use. It is suited for use in areas where freezing weather conditions may be found.
11. No. Intermediate hydrants should be 500 feet (152 m) apart.
12. All of the following are correct:
    Any obstruction(s) near the hydrant
    Outlets face the proper direction and have proper clearance
    Inspect for damage
    Condition of paint
13. Water supply officer

CHAPTER 4

1. False. The principal reason for conducting fire flow tests is to determine the rate of water flow available for fire fighting at various locations within the distribution system.
2. False. A pitot tube is used to record the velocity pressure of water being discharged from an outlet.
3. True
4. False. The hydrant should drain completely when closed.
5. D
6. B
7. D
8. A
9. A and D
10. C
11. *U.S.*
    Flow in gpm = $29.83 \times C \times d^2 \times \sqrt{P}$
    Where: 29.83 = A constant
    C = coefficient of discharge
    d = actual diameter of the outlet in inches
    P = pressure in psi as read by the pitot tube

    Metric:
    Flow in L/min = $0.0667766 \times C \times d^2 \times \sqrt{P}$
    Where: 0.0667766 = a constant
    C = coefficient of discharge
    d = actual diameter of the outlet in millimeters
    P = pressure in kPa as read by the pitot tube
12. 1,547 gpm (6 147 L/min)
13. All are correct:
    Does the water flow meet the needed fire flow for the occupancy?
    Is the available system flow adequate to provide sprinkler system demands in the area?
    Where will water system improvements be needed?
14. It is a correction factor that takes into account the amount of friction loss that occurs in the hydrant based on the construction of the hydrant.

CHAPTER 5

1. False. A minimum of 24 inches (610 mm) should be maintained all around the strainer.
2. False. If the flow rate is sufficient it may be possible to construct a dam, a rainwater basin, or a draft access basin to make the site acceptable for drafting.
3. A
4. D
5. C
6. 24 inches (610 mm)
7. Piped systems
8. *U.S.*:
   Q = A x V so, Q = (15 x 2) (8) = 240 ft^3/min
   240 ft^3/min x 7.5 gallons/ft^3 = *1,800 gpm*

   Metric:
   Q = (4.57 x .61) (2.43) = 6.77 m^3/min
   6.77 m^3/min x 1 000 L^3/min = 6 770 L/min
9. All of the following are correct:
   Inability to reach water with pumper

Wet or soft ground approaches
Inadequate depth for draft
Silt and debris
Freezing weather
Drying up

10. *U.S.*:
$(\pi)(11)^2 \times 4.5$ feet x 7.5 gallons/ft^3 = 12,830 gallons
Metric:
$(\pi)(3.35)^2 \times 1.37$ m x 1 000 L/m^3 = 48 300 L
11. A body of water the size of a football field and 1 foot (0.3 m) deep will supply 1,000 gpm (3 785 L/min) for five hours.
12. Use of dry-barrel hydrants, barrels or plugs floated in the water prior to freezing
13. Install a dry hydrant.

CHAPTER 6

1. False. It is more common to start out at the maximum flow and decrease it as required.
2. False. The attack pumper determines quantity of water needed during size-up.
3. True
4. False. Adjustments are made at the supply pumper first.
5. True
6. D
7. C
8. A
9. B
10. Number of apparatus, amount of hose, a reliable source
11. All of the following are correct:
Lack of training
Excitement, noise, and frequent interruptions affect pump operator's thinking process
Length of hoseline is unknown
Flow is unknown
Use of multiple hoselines of different diameters makes figuring friction loss too complicated
Gauges are not correctly calibrated
12. Open an unused discharge until a solid stream of water is flowing.
13. A. Intake pressure drops below 20 psi (138 kPa)
B. Operating the hand throttle does not result in increased rpm's.
14. Connect one discharge from a gated wye with one discharge of a clappered siamese.
15. Supplemental pumping provides additional water to a pumper when the hydrant it is attached to is not supplying enough.
16. A. Proper planning
B. Proper training

CHAPTER 7

1. False. Water is 10 percent heavier than fuel oil. Water is 20 percent heavier than gasoline.
2. False. Pumps on fuel oil tankers are generally of too low a volume to be useful to the fire service.
3. False. The main disadvantage is that the sides of the tank collapse under the weight of hard suction hose.
4. True
5. False. Top fill is least desirable method because of the hazard that serious line reaction may cause to personnel.
6. True
7. C
8. C
9. B
10. D
11. A
12. A
13. All of the following are correct:
At the source to fill tankers
To unload tankers that have an inadequate pump size or no dump valve
To draft from a portable tank or nurse tanker at the unloading site
As a shuttle tanker (pumper-tanker)
As the attack pumper
14. All of the following are correct:
Over weight limits
Possess a slower safe speed
Trouble traversing bridges and narrow roads
Increased response time

15. All of the following are correct:
    Previous product hauled
    Contamination from previous use
    Conversion effect on acceleration, braking, handling, and safety due to baffling, and tank security
16. All of the following are correct:
    Adequate but reasonable tank capacity
    Adequate loading rate
    Adequate unloading rate
    Adequate vent capacity
    Observe manufacturer's GVW rating
    Good roadability
17. All of the following are correct:
    Overhead
    Direct tank inlet
    Threaded tank discharges
    Pump-to-tank piping
    Tank-to-pump piping
18. Across the middle of the tank
19. Draft
20. Portable pumps generally do not have a large enough capacity and therefore take too long to fill tankers.
21. All of the following are correct:
    Collapsible or folding style
    Round, floating collar type (self-supporting)
    Homemade, piece together
    Four corner post style
22. All of the following are correct:
    Hose bed compartments
    Hinged racks above the hose bed
    Enclosed compartments
    On the side of a pumper or tanker
23. All of the following are correct:
    Act as a safety valve in an open relay.
    When a hydrant has damaged threads, it may discharge into a portable tank. Then the pumper can draft the water out of the tank.
    To provide at least minimal water from low flow hydrants.
24. All of the following are correct:
    Narrow roads
    Long driveways
    Blind curves
    When approaching other apparatus
    Winding roads and hills
    During inclement weather conditions
    Limited apparatus and driver capability
25. Gravity dump and jet dump. The jet dump is generally the quickest.

Index

NOTES

NOTES

NOTES

NOTES

COMMENT SHEET

DATE ____________________ NAME ____

ADDRESS ____

ORGANIZATION REPRESENTED ____

CHAPTER TITLE ____ NUMBER ________

SECTION/PARAGRAPH/FIGURE ___________________________ PAGE ___________

1. Proposal (include proposed wording or identification of wording to be deleted), OR PROPOSED FIGURE:

2. Statement of Problem and Substantiation for Proposal:

RETURN TO: IFSTA Editor
Fire Protection Publications
Oklahoma State University
930 N. Willis
Stillwater, OK 74078-8045

SIGNATURE ____________________________

Use this sheet to make any suggestions, recommendations, or comments. We need your input to make the manuals as up to date as possible. Your help is appreciated. Use additional pages if necessary.